THE ABYSS OF TIME

THE ABYSS OF TIME

AN ARCHITECT'S HISTORY OF THE GOLDEN SECTION

Martin Hutchinson F.R.I.B.A.

Chocolate Tree Books

First published 2008

Published by Chocolate Tree Books, Eastbourne

Printed by Lightning Source

Publisher site: *www.chocolatetreebooks.com*
Support page: *www.chocolatetreebooks.com/abyss/support.html*

British Library Cataloguing in Publication Data

Hutchinson, Martin

The abyss of time : an architect's history of the golden section

1. Architecture - Composition, proportion, etc.
2. Golden section
3. Fibonacci numbers
4. Architecture, Ancient

I. Title

720.1'51272

ISBN-13: 9780955706813

Hutchinson, Martin (1903-1973)

190pp. (i-xii, 1-178) size 234 × 156mm

includes index, and 40 b&w illustrations

Foreword

Martin Hutchinson F.R.I.B.A.
(1903 – 1973)

Martin Hutchinson's father and grandfather were both professional artists. He showed talent for drawing and painting too, but took the advice of his father never under any circumstances to earn a living as an artist, and instead found work as an architectural assistant/draughtsman with a civil engineering firm, studying for his architectural degree at evening classes.

My father pursued his passion for painting and drawing throughout his life; a sketch pad and watercolour box would accompany him on every holiday, cycle or walking trip. He produced hundreds of fine "plein-air" works and had a particular talent for buildings and boats as well as landscape. As he gained experience he could render the most complex of detail with real economy of brushwork.

The contrast between these small works and his output after qualifying as an architect could not have been greater, for he designed some of the largest industrial buildings of his time including the Ford assembly plants at Dagenham and Halewood.

He found time for many other interests, including model making, carpentry, and sailing. He also patented six inventions, notably Viaspan, a portable road flyover system, and played and composed music despite being profoundly deaf. He was a profuse writer, both of prose and poetry, and had articles and letters on a wide range of subjects, serious and humorous, published in many journals and magazines.

His interest in, and subsequent study of, the Golden Section – along with his research on measurement systems – derived from his work as an architect and his observation as an artist. From this he produced an initial manuscript in 1970. When he died in 1973, the work was essentially complete, and it is now finally being set and published for the first time for release in 2008.

Matthew Hutchinson

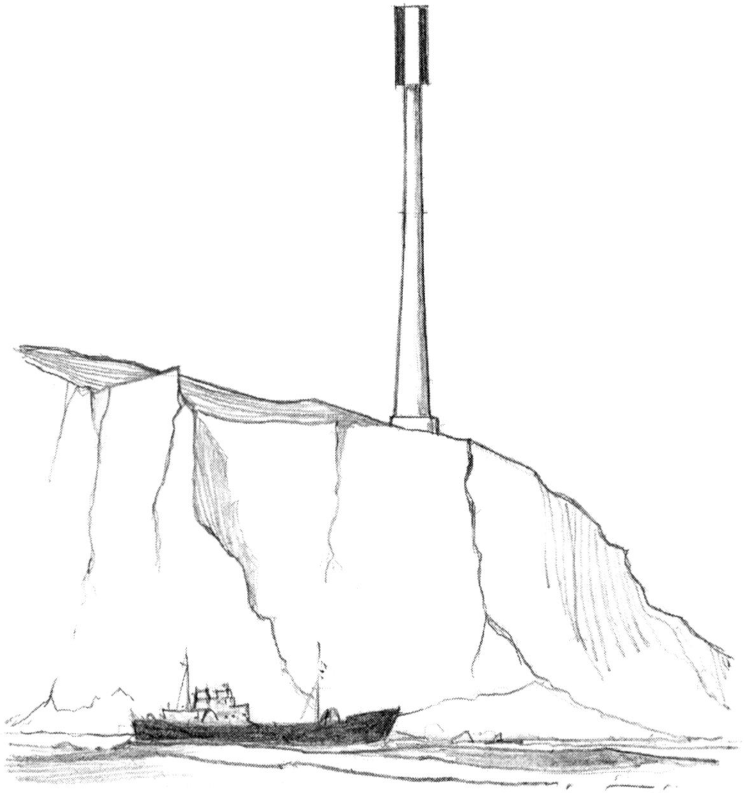

Navigation Beacon at Milford Haven

A modern example employing the Golden Section system of proportion.

The total height of the tower is 160' 0". The height ratio between that of the mark, the vertical portion of the shaft and its tapering lower part is in the proportion of 3:5:8

Introduction

Because this book deals with matters remote from everyday experience, yet for all that, intimately connected therewith. I must make my standpoint clear at the outset.

During the past fifty years or so we have acquired vast stores of knowledge yet we still find many things about us that raise questions to which there are no ready answers. The general tendency is to dismiss such uncomfortable neighbours from our minds altogether, or alternatively provide explanations that are a shade too facile.

This attitude of mind is matched and counterbalanced with another. We live in a credulous age, and all manner of practices, based on superstitions inherited from former accredited beliefs hold sway and in various degrees affect the outlook of the multitude. Astrology, palmistry and crystal-gazing are cases in point.

This treatise is much concerned with the concept of number, so out of this array of rife quackery I choose to pick upon one of the less pervasive, numerology, a pseudoscience that seeks to endow numbers in themselves with attributes, usually of a transcendental nature.

Numbers are, indeed, strange things, existential and non-existent simultaneously. Furthermore they can, like some animals, be coerced into the performance of strange tricks, but, as with animals, these serve only to detract from their true nature.

My attitude towards numbers is the direct antithesis of that implied by numerology; for I hold that it is mankind's invention and usage of them that imbues them with whatever properties they may possess.

Still on the subject of superstition, it must not be forgotten that new brands appear amongst us, born even of the very modes that seek to discredit them. The U.F.O. cult, the offspring of an uneasy alliance between science and fantasy, is one of these. With the best will in the world we seem to be fabricating other superstitions as well.

When the answer to a question generates problems still more intractable, this may denote a legitimate field of expanding enquiry. On the other hand, it could indicate that the wrong question has been put in the first place. As Sir Arthur Eddington once wisely said, "*It may be found in the end that many of the questions we pose are meaningless*" An instance of this concerns some widely accepted interpretations of the "red shift" These are of such a nature that the mind boggles, and I cannot discount the possibility that some of the conclusions that have been drawn will eventually assume their proper guise as good red herrings.

The reader will find that I have an entirely different attitude of mind towards folklore and mythology, because I am of the opinion that many an ancient tale, embodying tradition and race-memory, possesses some solid factual basis even if in its present guise it has departed far from its origin.

He will also notice that, in developing the arguments I often quote architectural examples. This for me is an occupational hazard, and I make no excuse, because in one form or another, works of architecture, as were once the poor, are always with us.

I have been at some pains to expand the theme in plain English, avoiding the esoteric and ambiguous terminology so much in fashion today; and the occasional blinding revelation of the obvious is inevitable because many of the guide-marks upon the road that I have chosen are so intermixed with the tenor of our daily lives that they excite little curiosity.

Sometimes the text diverges from the line of the main argument, but only when the discussion leads to a more illuminating view of the point at issue. For the rest, it is at once a study of little-known relationships and a search for keys to open doors leading into long-forgotten rooms.

Martin Hutchinson, 1973

Contents

Contents

The Abyss of Time

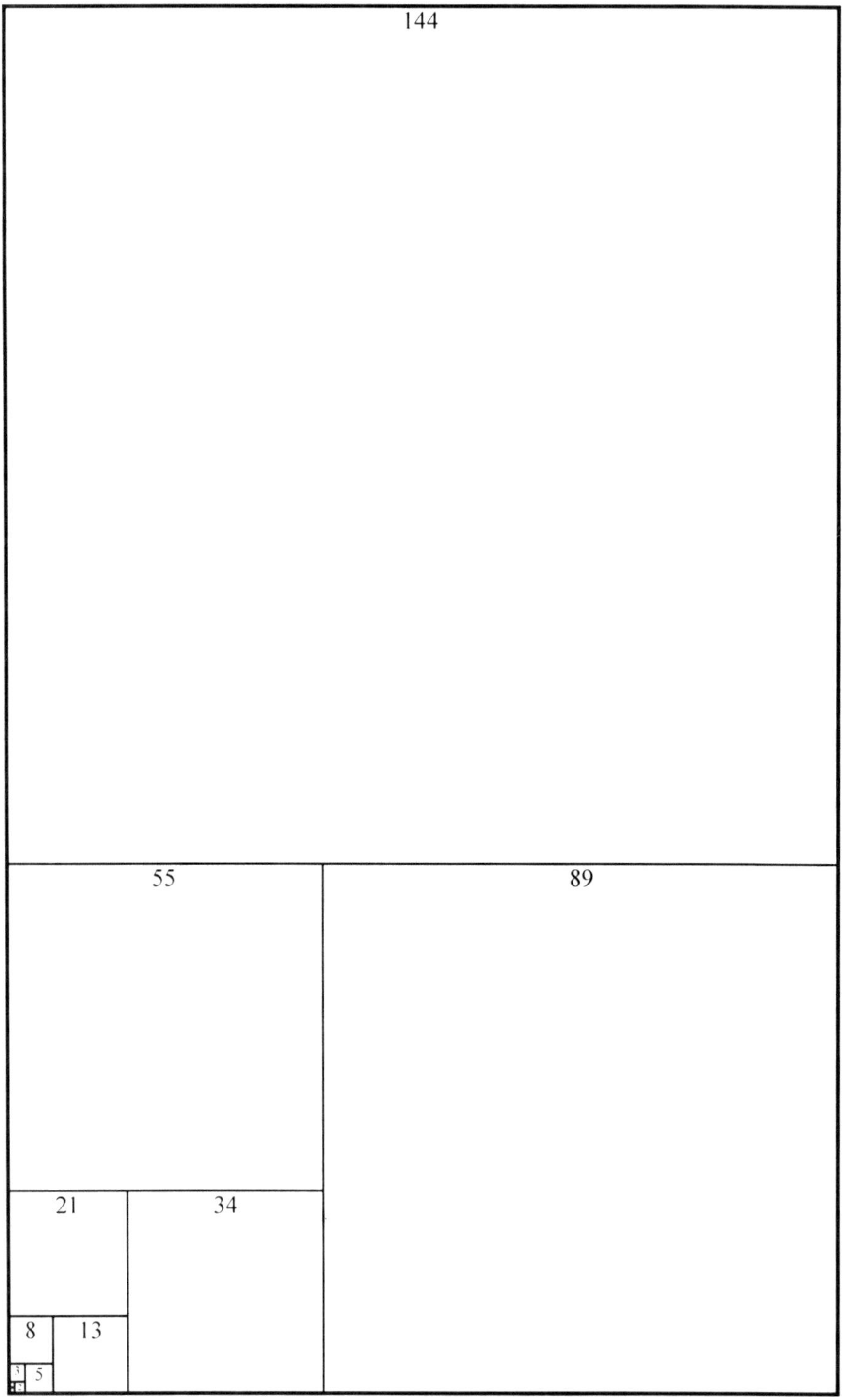

CHAPTER I

TO INTRODUCE THE THEME

ONE of the difficulties attendant on the preparation of this first very chapter lies in the nature of the subject matter to be treated. Had I chosen to write about agriculture, economics, bee-keeping or badminton, there would have been little need for an introduction save one to explain in what way such a treatise differed from others of its kind.

This book is of another character altogether, because it deals with matters the very names of which have little popular significance, but which I nevertheless believe to have had a profound influence in moulding thought, habits, and much of the action in times long out of mind, and consequently those which prevail today.

Another problem has been that of presenting my story in a manner to carry convincement of the reality of the past: that the unrolling tapestry of time comprises something much more than the series of archaic pictures implicit in the strictly "historical" view. Each of these long-gone moments were part of a vital present for those who lived within their ambit. The tapestry is indeed serial, but in a manner whereby each successive section is woven out of the fabric of its predecessors.

St. Paul, writing to the Corinthians, refers to the punishments meted out to certain antique transgressors, and says that "*these should serve as a warning to us,* ***we in whom history has received fulfilment.***" That statement was made at a time that we regard as belonging to the distant past, but nevertheless it is valid for us today, who bear the burden of two additional millennia. It is we, here and now, who have received the fulfilment of history, and in their own time our great grand children will be in like case, the while regarding us as dim figures lurking in the ancestral groves; Yet their actions and modes of thought will have been much influenced in ways they little suspect, by what we now think and do.

The development and expansion of comprehension can be likened to an inverted pyramid, its apex on the ground representing the dawn of self-consciousness, its base in the air, the living present, the field of thought and action in each successive era, ever broadening as time moves forward.

The first man to strap a nodule of flint to a wooden handle, or the curiously garbed mason of the Middle Ages, piling stone on perilous stone, did not look upon themselves in the way that we regard them, quaint and somehow unreal beings occupying a less than three-dimensional world. In their own minds they were in the van of progress. In their own time they represented contemporary man.

This image of an expanding pyramid balancing on its nose is apt in more ways than one. Firstly, it depicts a form of structure which is intrinsically unstable: the more it expands the greater the risk of collapse. In the earlier phases, it might have been possible for a wise man to have formed a reasonably correct view of his own ambience, as then known, but this is no longer so, for now the frontiers of knowledge, like the bounds of space, are receding at an ever increasing rate. Innumerable possessors of left hands know not what those with right hands, those in danger of knowing too much, are engaged in doing. There is indeed reason to suspect that a critical condition approaches, when top-heavy and beneath the burden of its own weight, the whole pyramidal fabric, our civilizations with it, will topple to ruin.

In another way, the analogy represents the influence of that powerful factor, tradition. Like Professor Hoyle's concept of continuous creation, tradition is ever renewing itself, despite sporadic attempts at suppression. These, self-defeating, because they lead to dead ends, are doomed to failure and it should be noted that in the last resort, even the periodically recurring revolt of a new generation against the modes of its predecessor is a traditional exercise in rejection.

This concept of a "pyramid of time" is shown in the diagram below:

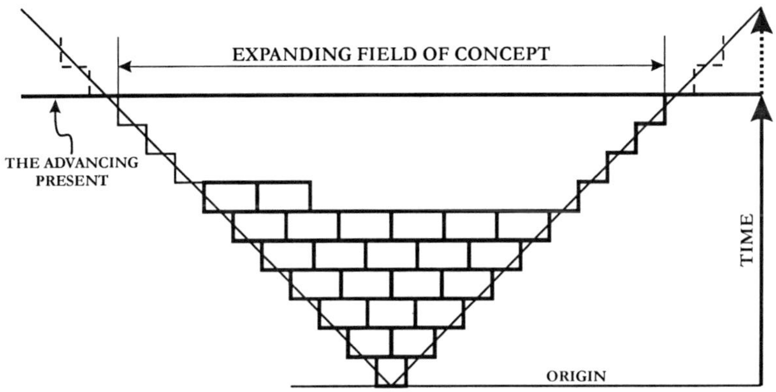

Figure 1

Each layer represents the elapse of the years, a period not necessarily an age, such as the Stone Age, the Bronze Age, the Middle Ages, or even the Victorian Age, which properly speaking are not ages at all, but the names we give to the broad aspect of cultures, occasionally coeval and usually confined within restricted ethnical or geographic limits.

The components of each layer form the foundations upon which its successor is built up, a long series extending back into the remote depths of time.

Some of these figurative building blocks have fallen into disuetude, but others have continued to exercise their clearly defined function as fundamental supports up to the present day.

There is a category comprising those, the functional influence of which has exerted influence over immense periods of time, sometimes strongly, at others feebly to the point of extinction, and it is the study of one of these, at present in little regard, which is the main concern of this book.

The influence to which I refer is a proportion, known as the Golden Section, the properties of which have exercised the imagination for many thousands of years. In its physical form it has been utilized in innumerable ways, and has been the object of conjecture and comment throughout the ages. It can be expressed by quoting a pair of numbers, and from this initial pair, a whole numerical series can be built up.

This series, the Fibonacci Numbers, was formulated in the 13th century, but there is very good reason to suppose that the numbers concerned were known at a very much earlier date, utilized in diverse ways and influenced choices to an extent that has determined the nature of many present day modes.

The book, within the limits of its own context, is an investigation, retreading the tortuous paths of the past in quest of derivations by reference to survivals, including artefacts and concepts, and even touching on myth and legend.

The evidence is of differing kinds, the first in which conclusions can be drawn from physical examples, the others concerned with concepts the origins of which, as often as not, are shrouded in obscurity.

One example in each category will suffice to summarise their characteristics.

(1) Physical Evidence

We can see and examine sickles of prehistoric manufacture, the blade in some cases shaped from a single stone, in others comprising small flints set in line in antler bone. We have evidence of the slow evolution, the change from stone to bronze, from bronze to iron, and from iron to steel blades, the like of which we still use. We can thus envisage the steps in the long process of development culminating at last in those complex machines, that towards the end of Summer, harvest our fields today.

(2) The Related Concept

The principles involved in a prehistoric dew-pond are the same as those in the modern steam-condenser, and involve the design of a heat exchange system to reduce water vapour to water. The appreciation of the principles, separated by a vast stretch of time, is the same in both cases.

(3) Concept of Unknown Origin

This is of a different nature altogether. In this example, it is displayed in a concept which seems to lie upon the frontiers of the abstract. It is a system of measurement, the origins of which are quite forgotten, the archaic measure for wine, in which 8 gallons constitutes a "pipe", and just over 13 pipes defines the capacity of the cask.

The increment appears to be irrational. A "pipe" at eight gallons sounds sensible enough, but the choice of such an incompatible and unruly number as 13 for the next is seemingly beyond the bounds of reason. Yet the choice was made, and the mode persisted, so one must infer that at its inception it rested upon some sort of logical basis. As I shall presently show, there are other examples of this nature which remain puzzles until some pattern of unification and relationship is brought to light.

4. Unrecognised Survivals

There is yet a fourth category, that which is found in the nature of concepts that have survived into the present, but which because of their familiarity, excite little attention. An instance of this kind of survival is exemplified in the ordinary "foolscap" paper such as that upon which the manuscript for this book has been written. The dimensions are 8" × 13". Now the shape of this paper is an example of the Golden Section, and the two-dimensional figures are part of the numerical series mentioned previously.

It will be observed that these dimensional figures are practical equivalents of the numerical values involved in the rather absurd wine measure described above. Another even more striking relationship becomes apparent when we take into account the length of the link in the Surveyors chain, a little less than 8 inches, and the length of the common early Mediaeval building foot, at slightly more than 13 inches (13+ again). In this case the survivor is the link. The ancient foot has long fallen out of fashion.

This book is the tale of an endeavour to discover and establish the relationships that examples of this kind, flotsam from the past cast upon the contemporary shore, bear to one another. To mix my metaphors a little further, it is an attempt at the disentanglement of a tenuous thread in a skein of causation, reaching far back into antiquity, to an era so totally and wholly forgotten, that no other hint or record of its one-time existence remains.

1	1	1
2	1	1
3	2	10
4	3	11
5	5	101
6	8	1000
7	13	1101
8	21	1 0101
9	34	10 0010
10	55	11 0111
11	89	101 1001
12	144	1001 0000
13	233	1110 1001
14	377	1 0111 1001
15	610	10 0110 0010
16	987	11 1101 1011
17	1 597	110 0011 1101
18	2 584	1010 0001 1000
19	4 181	1 0000 0101 0101
20	6 765	...
21	10 946	...
22	17 711	...
23	28 657	...
24	46 368	...
25	75 025	...
26	121 393	...
27	196 418	...
28	317 811	...
29	514 229	...
30	832 040	...
31	1 346 269	...
32	2 178 309	...
33	3 524 578	...
34	5 702 887	...
35	9 227 465	...
36	14 930 352	...
37	24 157 817	...
38	39 088 169	...
39	63 245 986	...
40	102 334 155	...
41	165 580 141	...
42	267 914 296	...
43	433 494 437	...
44	701 408 733	...
45	1 134 903 170	...
46	1 836 311 903	...
47	2 971 215 073	...
48	4 807 526 976	...
49	7 778 742 049	...
50	12 586 269 025	...

Fibonacci Series, in decimal and binary

CHAPTER II

THE DEVELOPMENT OF NUMERATION

There are signs that in the interests of technological innovation, the numerals that we know, will eventually become obsolete and not long hence relegated to the status of back-numbers. The days of numerals are numbered.

However, much of this book is concerned with them. As they play such an important part, it will be well, at the outset, to consider how they came into being, and the changes that have come about in their long history.

The tale of mankind's development from the beginning is like an account of an immensely long journey, in which the last five minutes is described with a fair degree of accuracy. Before this the story becomes ever more vague to a point where the fog of incomprehension becomes too thick to afford a view of the early stages, let alone the commencement.

There are, however, two reasonable assumptions that can be made. Firstly that the observation of the difference between the one and the many must have been coeval with the dawn of Man's self-consciousness, the creature's recognition of his own existence in relation to an objective universe around him, and, secondly, that this must have soon led to a rudimentary concept of number.

This and the process of counting, one bound up with the other, was the seed and origin of the mathematical sciences that play such a large part in our present-day affairs.

A synopsis of development somewhat broad in outline is given below.

The later stages were not necessarily successive, for it is known that the ancient Mesopotamian civilizations possessed a self-contained system of numeration (stage "f") thousands of years before Greeks and Romans involved themselves in the difficulties arising from the use of alphabetical symbols to represent the numbers as well.

(a) **Signs and sounds** leading to speech and counting (i.e. numeration in the form of words, audible or not as the case might be.)

(b) **Numbers represented physically**, perhaps by counting fingers and toes, but also by assembling physical objects such as stones. ("Calculus" is the Latin for "pebble")

(c) **Proto-literacy**, by means of ideograms.

(d) **Literacy**, involving the use of pictograms and hieroglyphics, symbols representing phonetic equivalents and in the end a system representing vowels and consonants.

(e) **Numeration** by means of these last, thus put to a duplicate use.

(f) **Adoption of systems of true numerals**, symbols in their own right.

(g) **The facility for manipulating these to serve practical ends (arithmetic)** and the extension of this together with the utilization of alphabetical symbols in pursuit of abstract speculation as well (algebraic); in short the process of mathematical calculation.

It is to this that we owe the invention of our own sophisticated methods of communication, telephone, radio, and television, and, it is amusing to consider that this ultimate so far, is but a more elaborate way of providing the same service as stage **(d)**, the paintings in the caves of Altamira, or at a much later date, the stained glass windows of the Middle Ages.

II: Development of Numeration

As in other instances, the path of development was erratic, even in its earliest manifestations. We have a great deal of evidence which seems to indicate that the concept of numeration sometimes preceded literacy, pieces of bone twenty or thirty thousand years old, inscribed with groups of notches with accented intervals, long antedating any known symbols representing speech as such. As mentioned above, the inhabitants of Sumer had a true numerical system, but the Hellenic peoples at a much later date used the rather clumsy device of grouping the letters of the alphabet to represent a decimal order. The Romans, always practically minded, improved on this to a certain extent, by reducing the number of alphabetical equivalents, and bequeathed this system to Western Europe.

Both of these systems would work very well for the purpose of counting, but a natural intractability would debar them from use for purposes of calculation. If the reader finds this statement too sweeping, I suggest that, *without* recourse to "Arabic" equivalents, he works out this simple example:

$$\frac{\mathbf{XXIII}}{\mathbf{CVI}} = \ldots\ ?$$

Our own current numerical system is believed to have originated in India, some five or six centuries B.C., and comprised symbols that are the ancestors of the so-called "Arabic" numerals.

It was not introduced into Western Europe until Medieval times, and indeed the archaic system of Ancient Rome persisted until much later, and still survives. Even now, a Roman, resuscitated from the time of Marcus Aurelius could read many of our clocks once their function had been explained to him.

In the province of numbers and their use for the purposes of calculation, some cultures were far in advance of others. When the Sumerian sages, in the intervals between spells of practical work, grappled with abstract problems, the majority of their contemporaries elsewhere were still engaged in huntin', fishin' and (albeit less noisily than now) shootin' – for a bare existence.

The Abyss of Time

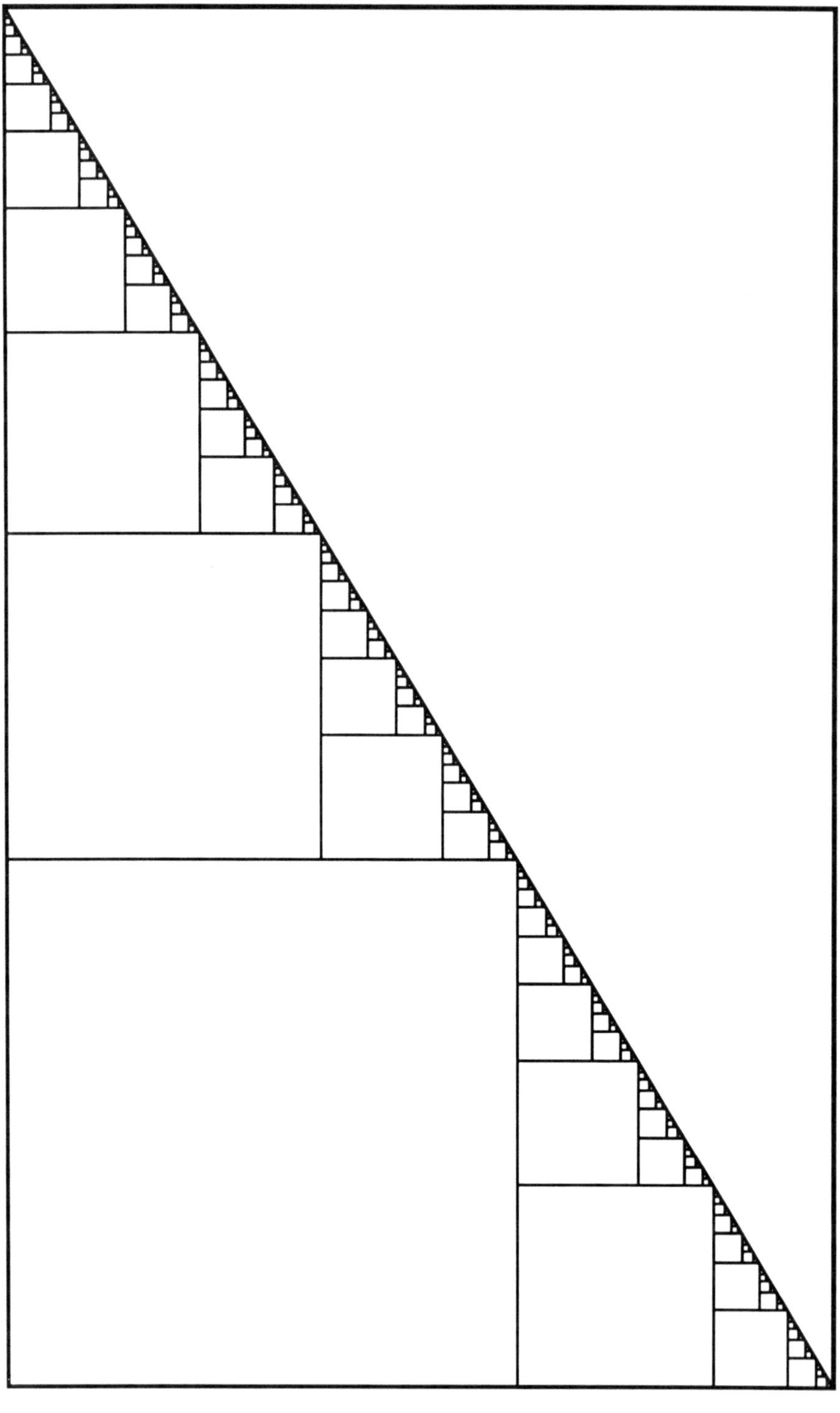

CHAPTER III

THE GOLDEN SECTION AND THE FIBONACCI NUMBERS

The proportional ratio, and the numerical series referred to in Chapter 2 are known respectively as the **Golden Section** and the **Fibonacci numbers**. These concepts have been the subject of speculation for a very long time. They have received so much attention that it would be fatuous to expand the theme unless there was something new to say.

For the benefit of the reader who is unfamiliar with either, I shall devote this short chapter to a general description.

(1) The Golden Section

Euclid defined the Golden Section thus:

> "A length so divided into two parts that the proportion of the shorter length to that of the longer is the same as the proportion of the longer length to the whole"

Euclid gave this definition some 2,300 years ago. This does not mean that he invented it, because there is ample evidence that it was well-known thousands of years before his birth.

It may sound a little complicated, but it is really very simple.

If we take a length of, say, 21", and divide it into two parts, one 8" long and one 13" long, the proportionate relationship between 8" and 13" is nearly the same as that between 13" and 21". This correspondence is not quite exact, but the slight inaccuracy, a fascinating subject in itself, will be dealt with later.

To aid visualisation, the proposition is shown in diagram form:

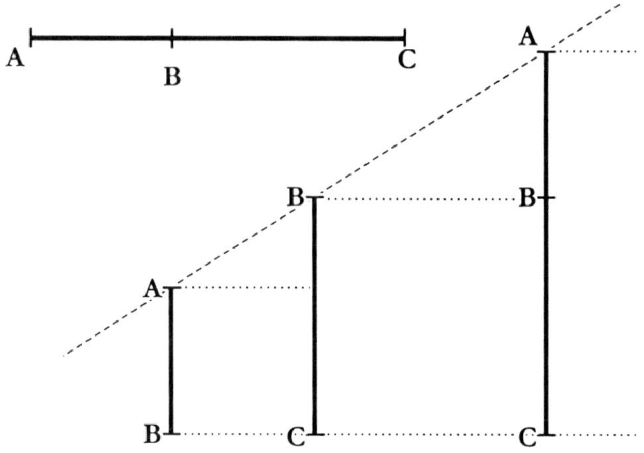

The line AC is cut at B. The proportion between the two parts, AB:BC, is the same as that between BC and the total length AC

Figure 2

We can manipulate the two component parts to construct a right-angled triangle ABC, and putting two of these together we get a rectangle ABC(B_2).

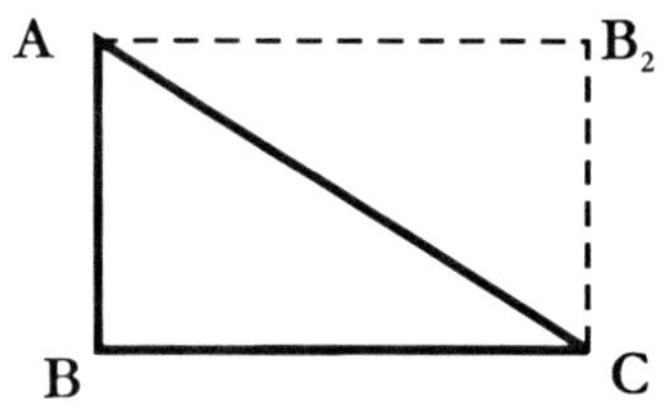

Figure 3

This is a Golden Section rectangle and as already mentioned it is the shape of a piece of English foolscap.

This proportion with a ratio of 1 to ~1.618 has held a particular fascination for artists and designers since very early times. It is found in pictures by Seurat, by Leonardo and Mantegna to mention but three in recessive order of date. Earlier, it was in use as the basic modular for the design of many of the great Mediaeval buildings.

The Greeks used it as a regulating proportion in the designs for some of their temples, but one of the most remarkable and beautiful extant examples and by far the earliest is the bas-relief slate palette recording the exploits of King Narmer, the first ruler of an united Egypt some 5,000 years ago. This is described more fully in Appendix A, which also analyses the basis of the design.

In all these cases it appears to have been used consciously, but in others, and quite often, it appears to stem from the subconscious.

Some years ago I was sketching with a young friend. He painted a street scene in sepia monochrome, except for the accent point and focus of the composition which was a scarlet blob representing a pillar-box. The paper was not a "Golden rectangle" in shape, but later we divided the sides into Golden Section proportions horizontally and vertically, thus:-

Figure 4

Point "A", where the ordinates intersect, was the position of the pillar-box. My friend never sold it, but a similar effort by the American artist Barnet Newman was substantially rewarded.

He covered a very large panel with an orange ground and indicated the Golden Section with a single vertical carmine stripe. This masterpiece induced a purchaser to part with $26,000.

J.M.W. Turner, surely one of the greatest British painters, has at times been criticised on account of the allegedly uncouth compositions and profiles that obtain in some of his pictures. It should be remarked that a denigration of this kind is of its own nature subjective, and like all value-judgements, itself subject to criticism.

Whatever the issue, Turner, consciously or unconsciously made use of the Golden Section.

A magnificent early picture "Fishermen upon a Lee Shore" when analysed in the fashion mentioned above is a telling example, because the main focus has been re-echoed no less than three times, providing a Golden rectangle within the picture itself.

A Romano-British wine jar, before me as I write, measures 10" high × 6 ¼" diameter, 1: 1.6 again. The proportion could have been chosen instinctively, or deliberately. It is but one of the many examples that I have found within the four walls of my home.

To return to the Golden Section rectangle, the angles that the diagonal makes with the sides are approximately ~58° 17' and ~31°43' (expressed in decimals, ~58.28° and ~31.72°) respectively; *

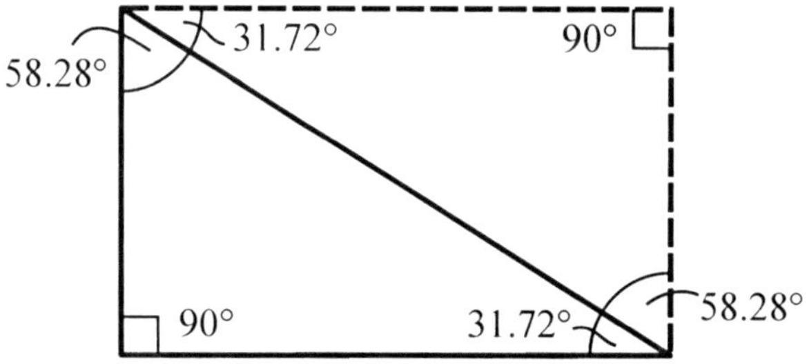

Figure 5

Dr. B.G. Morgan has made a deep study of the properties of the Mediaeval mason's square. This will receive more detailed comment at a later stage, but for the moment let it suffice to say that the characteristics of a 60° angle, and one of ~58°17' are both implicit in the conformation of the instrument.

It is clear that the proportion, and the shape that it engenders has been used throughout the ages, consciously or subconsciously, in the realms of architecture, painting, and decoration, as well as utilitarian design. One odd fact emerges. Although it has been so often employed in the composition of paintings, the actual overall shape of panels or canvases rarely conforms. In fact the only example that I am familiar with is a relatively unknown landscape by the English painter, R.P. Bonington.

---

* For all subsequently-mentioned angles, approximations will be assumed.

(2) The Fibonacci Series

The Fibonacci series, or Fibonacci numbers, named after their formulator, the great 13th Century mathematician and philosopher, is a numerical series of integers with a strong affinity with the Golden Section. It also possesses some peculiarities, of which more will be said at a later juncture.

It commences at unity but only exhibits the affinity from 3 upwards, and runs 1, 1, 2, 3, 5, 8, 13, 21, 34, 55, 89, 144, 233, 377, … and so on. It will be seen that every term is obtained by adding the two preceding terms. It will also be observed that from 3 upwards the proportion of one term to the next is always closely constant. It is in fact a numerical specification of the Golden Section.

Take any four at random and you will find the results to be alike within very close limits:

(a) 3:5 as 5:8 (5/3 = 1.6666' , 8/5 = 1.60)

(b) 13:21 as 21:34 (21/13 = 1.615…, 34/21 = 1.619…)

(c) 34:55 as 55:89 (55/34 = 1.617…, 89/55 = 1.618'18'…)

However, it will be noticed that the examples demonstrate a built-in inaccuracy. The figures do not quite agree. In the first the aberration is in the order of 1/25, but in the third it has reduced itself to 1/3026. The higher up the series, the smaller the error, until presumably at infinity, it disappears altogether.

When this level is reached, numbers as such must disappear, and if these symbols vanish, so must those other symbols, words.

We are thus doubly bereft. The means of formulating a conjecture and the adumbration of an answer seem to be beyond our grasp.

Nevertheless, I have made an effort to deal with this thorny problem in a later chapter. This one serves but to describe the phenomena and to emphasize the close relationship between them.

The Abyss of Time

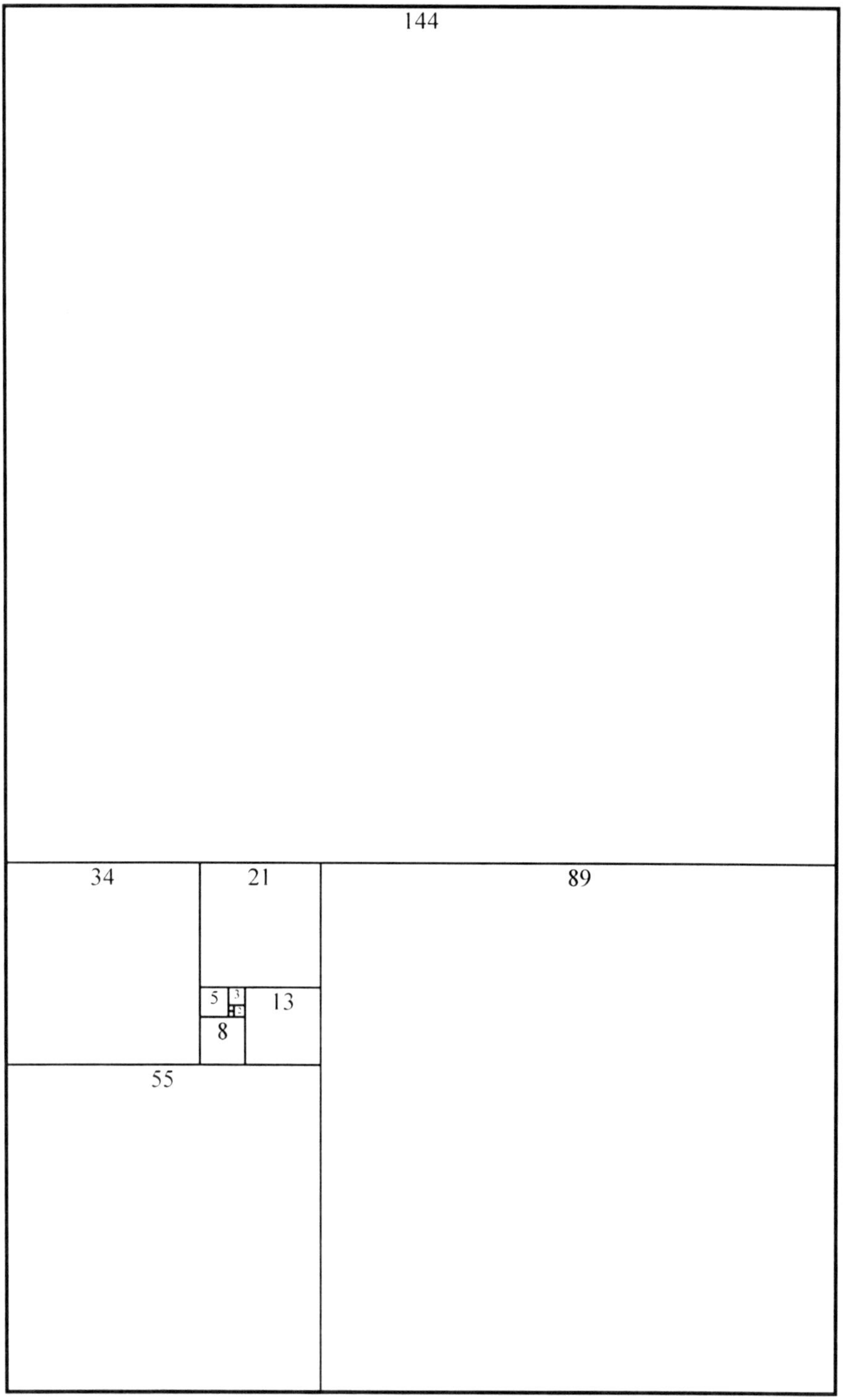

CHAPTER IV

IN PURSUIT OF THE FOOT AND OTHER QUARRY

(1)

The development of the theme and an assessment of the evidence must wait awhile for it is now necessary first to consider some of the steps that led to eventual conclusions.

Observation and enquiry in the realm of mensuration indicated that the length of certain measuring units were related to one another in the proportion of 1: 1.62. It also appeared that the number of measuring units used in varying kinds of tasks conformed to the "Fibonacci" numbers. The first statement is beyond dispute, but the second depends upon the provenance of the unit by which the examples have been measured, to wit, the foot or the inch.

If a prehistoric earthwork measures, say 144 feet in diameter, there can be no cause for jubilation at this coincidence with a Fibonacci number, unless the current foot which defines the measurement can be shown to be as old, or older than the earthwork. In short, the proportional characteristic could be justified, but there was some doubt as to the cogency of the numerical equivalents. An investigation into the antecedence of foot and inch was clearly indicated.

The beginning of this study was not very hopeful. Little literature on the subject could be found and what there was, although of expert origin, appeared to be inconclusive, or veiled in obscurity. The writers, scholars, were writing for scholars, not general readers. Such need the sort of synopsis that prefaced the old-fashioned magazine serials, commencing "The Story So Far", and concluding "Now Read On".

The first lucky break occurred when I read about the "will" of Henry VI, the charter dated 1446 for the building of King's College Chapel at Cambridge. This specifies the main measurements to be adopted for the various parts of the building, given in feet.

When the building is measured today, using the current foot, substantially the same results are obtained.

In the 15th Century, all manner of measuring units were in vogue, including the ell, the cloth-yard, the long foot and other variants. More than a hundred years had to pass before some measure of standardisation was introduced, the precursor of the Imperial system that we are just about to abandon.

This employment of the current foot for work under Royal patronage proves that 500 years ago it was pre-eminent amongst its fellows, well-established, and with some tradition behind it. Somewhat later, I found that the common Mediaeval foot of 13.08" fell into disuse in the 12th Century and that in the reign of Edward I, the length of the foot was very close to that which we know today. This took me back some distance in time, but by no means far enough. There seemed to be a dead-end. The British Museum was unable to help beyond suggesting consultation with the Science Museum.

Here I was rewarded with the observation that the current foot derived from the Roman foot of 11.72". Indeed, there is little to choose between them. This was more hopeful – a further 1200 years at one jump – but still unsatisfactory, because my interest was involved with things that were much older still.

Some time later, visiting the British Museum for another purpose, in the entrance lobby to the Edward VII Gallery, I chanced on a little display of maps and ancient methods of land surveying, including a model of an Egyptian measuring rod, one royal cubit in length. This measure, some 5-6000 years old, in its own life-time was sub-divided in different ways, but in this instance the method indicated division into **21** equal spaces, digits.

The Royal cubit is 20.62 inches long, so the digit is 20.62" / 21 = 0.982", and 12 of these measure 11.784" which is, of course, just over 0.98 feet.

If we regard the digit as a short inch, the reduced foot of 12 short inches is less than 2% shorter than the current foot. The Royal cubit is reputed to date from the pre-dynastic era, that is, B.C. 3000 or thereabouts, but just as we found the current foot well-established 500 years ago, we can argue that its component ancestor, the digit, or short inch, could possess an equally remote ancestry of its own.

It may be contended that there is no direct evidence of the "short" inch or digit being used in multiples of twelve. This is conceded, but it should be noted that other ancient equivalents, such as the Oscan and Drusic foot both approximate, if not so closely, to the same length as the conjectural foot described above.

The inch and its corollary the foot therefore appear to be at least 5000 years old. Additional evidence may yet be found to prove what I strongly suspect – that these units are much older still.

(2)

The investigation that has just been described was accompanied by speculation on the interrelationship between units of measurement, and measuring devices. The latter, or occasion, have evolved their own character, and by so doing have initiated a new and larger basic unit.

For example, that comparatively late arrival, the fathom, comprises six feet but is never so divided. Soundings on present day charts are given in whole fathoms, or fractions thereof.

The reason is simple to understand. As soon as ships attained a reasonable size, the all-important question, in any wind or weather would be "How much water have I got under my keel?" If the answer was in inches, or say a foot or two, the safety margin would be far too narrow, so Tiphus or Ulysses plumped for a measure incapable of ambiguous interpretation, and one that left nothing to chance. The sounding rod of reasonable and handleable length would have been evolved, and later when ships grew larger the lead-line, still divided into fathoms, each identified by a distinguishing mark.

Then there was the ell. The length of this varied widely, each version evolved, no doubt, for its own honest purpose. However, this ambiguity could have proved a boon to the trader – short measure and a ready-made exculpation at one and the same time.

Another large unit was the old French "canne" of just over 78 inches, equal to 6 old English building feet. This measure is found in Gallic work of the Middle Ages, and also at a much earlier date. The modern translation is "cane; walking-stick", and I have little doubt that this meaning stems from "staff" or "rod". In other words, the canne was not only a name for a unit of measurement, but for the actual measuring device as well.

Something of the same sort can be seen in the hands of the peripatetic official who has the job of testing the gauge of railway lines. It is a beautiful precision instrument of polished hardwood and brass, made for the particular purpose. It does not concern itself with centimetres, inches, feet or yards. It's sole job is to report whether the prescribed distance between the rails is correct or not and it provides the information immediately.

This leads me towards the conjecture that a device of a similar character was in use a very long time ago. Standard railway gauge is 56 ½", but quite often we find that a dimension comprising 55" or thereabouts was used for various purposes in early historic as well as prehistoric times.

This measure, if such there is, is long obsolete and has no name, but in its physical aspect it would possess the advantages of portability. It would also serve other needs – helping the weary, or prodding the reluctant ox. There is some confirmation of this concept from another source. Those peoples, who long ago made use of the "Megalithic yard" appear to have utilised it in multiples of 2 ½, that is a "preferred" dimension of just over 6' 10". Immediately, one visualizes a wand or rod of this length, handy for defining measurements of all kinds, for setting out the bounds of an enclosure or for establishing the width of the gap in an earthen wall to be barred once the herds were safely inside.

This concept of a staff incorporating a fixed measure has other connotations. The Long Man of Wilmington, the "Dolman", holds two, so it is not unreasonable to suppose that the club of Heracles had another function in addition to its use as a weapon. He who wielded the rod would always have been regarded as possessed of power above and beyond the ordinary, as indeed his achievements would show.

This, the measuring staff, and the wand to this day carried by the functionary, even the Royal Sceptre, can be regarded as at one with the rod of the magician. All are symbolic "wands of power".

(3)

Amongst those problems that admit of no solution we may ask the question as to which came first, the chicken-foot or the inch-egg. For my part, I incline to the opinion that the larger units were built up as an assembly of the smaller ones, just as one builds a wall from bricks. The reverse is a destructive operation.

At first this may appear to contradict what was said in the previous section, but this is not necessarily so.

Just as in these days, whereas a millimetre is essential to the watchmaker, it is a poor assistant to the designer of docks and dams. It must also be remarked that in the times when the concept of mensuration dawned, requirements were much less complicated than now. A relatively small number of assemblies of inches or feet would meet all practical demands.

The smallest units would be found convenient up to a certain limit; in the manufacture say of articles for domestic use, or weapons.

However, when larger scale work was to be undertaken, the use of the "digit" or "inch" would involve large and intractable numbers, perhaps beyond the comprehension of the practitioners. Such a unit would be quite unsuitable for measuring the length of a plot, or setting out the radius of an earthwork such as Silbury Hill. So they would be aggregated into larger units such as the foot, the cubit, the yard, the ell and the canne.

When distances, in contra-distinction to dimensions were involved, these units in their turn, would be found to be too small and they would be reorganised and grouped in the form of the stallion, the furlong, the mile and the league.

This enlarging process is still at work. Concepts of distance are ever achieving greater magnitudes. The astronomical unit (AU), the mean distance between the Earth and the Sun, retains the principle of linear measure, but is far too small to be usefully effective in dealing with the great distances involved in present-day cosmological enquiry. In this field, linear measure has been abandoned altogether, and the light-year invokes velocity alone. However, as the frontiers of the universe recede, even this system is beginning to feel the strain.

This synthesis, this building up of large things from small ones is logical, just plain common sense, but it seems strange that this country, in this age, should see fit to push it beyond sensible limits. For as I understand it, in the near future, the basis of all linear measurement is to be the millimetre, a unit very difficult to appraise accurately. Most people can visualize the inch and the foot, or for that matter, the centimetre. They have a certain tactile quality, and one can imagine oneself seizing one, betwixt finger and thumb. The millimetre lacks this friendly relationship. Without a handy measure one could well be in error to the extent of 50% or more. One needs good eye-sight and a pair of tweezers to pick up a millimetre. Yet, henceforth, the largest things that we make will have to be conceived in terms of this tiny unit. The super tanker or container ships will no longer measure say 1200 feet between perpendiculars, but between mould-loft and building-berth will evolve to 360,000 mm.

Eventually, I suppose, we shall be faced with a decimal day of 10 hours, each hour comprising a hundred minutes and a thousand seconds, and to conform with linear measure, a clock dial on which seconds alone are indicated.

CHAPTER V

THE FINDINGS

This is a resume of the evidence supporting my theory that there was once a compatibility of thought underlying the choice of numbers selected for many different purposes. The emphasis is therefore on the numerical aspect, as distinct from the proportional characteristics that this represents.

For two reasons, this catalogue is far from being exhaustive. In the first place, a complete list would be so extensive as to allow boredom to usurp the throne of interest.

Secondly, it can be no study in depth, for, so far, it seems that I have exposed but the summit of the visible tip of the iceberg. New facts and clues continue to appear and call for attention, singly or in groups, sometimes at intervals of a few days, sometimes weeks. Even as I write, a family of six has presented itself. These many years, the members, to my shame have waited on my doorstep clamouring for recognition.

Some preliminary discussions have evoked criticism. People maintain that I am guilty of selective bias, and that credibility would be enhanced by general conformity.

The objection to this argument is, that if everything conformed, we would be living in a different world. All our methods of numerical assessment, our monetary and measurement systems would not be as we know them. Complete conformity would imply universal acceptance, hence the content of this investigation would have no meaning whatever.

Other sceptics object on the grounds of the heterogeneous nature of the sources quoted. I take the opposing view, that if fields, widely differing, offer evidence or traces of evidence of a similar character, this points towards some common basis.

If, at some multi-racial meeting members of different ethnic groups should each express similar opinions, this could be taken to imply that there was a consensus of thought and purpose.

There are also those who dismiss the matter as a wagon-load of coincidences. I cannot agree. As the next few pages will demonstrate, if this were so, the proverbial long arm of coincidence would seem to have been stretched far beyond its elastic limit.

I had long known of the existence of the Golden Section, the proportion 1: 1.618... , and had used it occasionally in my own work. I also knew that the golden rectangle could be set out with reasonable accuracy by drawing the sides in the proportion of 8:13 or 13:21, and that these are the functions of the cotangent of approximately 58½°.

(1) Eight, Thirteen, Twenty-One, Thirty-Four

Reading an article by Flinders Petrie I noticed that three archaic measuring units, the link, the old English building foot and the Royal Cubit, all measured in current inches, were defined so:-

Link	7.92
O.E.F.	13.08
Cubit	20.62

These were close enough to the Fibonacci numbers to arouse interest. Shortly after this I learnt that Professor Alexander Thom had isolated what he describes as the Megalithic yard, 2.72 current feet = 32.64"

Here then were four mensural units, each comprising numbers of component divisions that are very close to the Fibonacci whole numbers.

V: Findings

If we substitute the reduced inch mentioned previously we get a somewhat closer correspondence:-

	Length	Fibonacci Number
Link	8.08	8
E. F.	13.34	13
Cubit	21	21
Meg. Yard	33.3	34

It is probable that the two first items are derived from a Sumerian measure of about 39½" (a 5000 year old predecessor of the metre) i.e. the O.E.F. is one third of this and the link one fifth. Then we divide this measure by 13 we get ~3; when by 5, ~8; when by 8, ~5; and when by 3, ~13. So we have an inverse series of reciprocals:

Fibonacci	3	5	8	13
Reciprocal	1/13	1/8	1/5	1/3

The 39 ½ (or whole figure 40) unit is thus a non-conformist appearing among the orthodox. Although it does not appear in the Fibonacci series, it would appear to be related to it.

There are therefore three known and one hypothetical measuring units the lengths of which, in inches, all conform with the numerical series. Moreover, we find that they are all in the same proportion to one another. As the length of the link is to the Mediaeval foot, so is the latter to the Royal Cubit, and the Royal Cubit to the Megalithic yard. If we postulate the 55" measure mentioned in Chapter IV, the correspondence is extended a step further.

If the series had been a straight-line arithmetical one, say 6 – 12 – 24 – 48, there would he nothing very remarkable about this correspondence. But it is not, it is curvilinear, and to my mind the fact that the point reference for three known, one hypothetical, and one postulated measurement, all lie on this curve is something more than merely coincidence.

Other examples of the use of measurements of 34 feet and 39/40 feet are cited in Chapter XII and there is at least one instance, perhaps coincidental but possibly one amongst many, of the utilization of this particular number as the basis of design.

This is an Egyptian 12th Dynasty necklet in the form of a torque comprising 33 pendants and two terminals. There are therefore 34 spaces separating them.

(2) Fifty-Five

(a)

Every ancient artefact so far recovered has been found to have been made to serve a simple purpose, so we tend to confer a like simplicity of outlook on maker and user.

If the discoveries had included complicated mechanisms such as watches, typewriters or tape-recorders, we could no doubt take a different view.

I believe that this subconscious equation of simplicity of use with that of intellectual achievement is entirely erroneous.

There are many examples in which prehistoric polished stone of superb workmanship show that their makers had an eye for an extreme nicety of proportion and shape. This can be taken as an index to the predilections of the culture that produced and used them; for it argues a demand for excellence in things that would have served their purpose equally well had they been cruder and much less shapely.

This argument is borne out by the artefacts which are the subject of this section.

Just before World War II, Dr. Karl Absolon working near Věstonice in Moravia made some unprecedented discoveries. Amongst these there were a number of man-made objects since proved by radio carbon technique to be at least 20,000 and possibly 25,000 years old, that is, they were of a time when the ice which had covered much of Central Europe and all the North for so long, was at last in retreat.

V: Findings

Three of these artefacts concern this enquiry, the first being a tiny likeness of a human head carved out of mammoth bone.

Other likenesses of a similar date have been found elsewhere, roughly modelled out of clay and barely recognisable for what they are. This is of an entirely different order. It appears to be the portrait of an extremely good-looking young woman, and the artist, equipped with a masterly technique, has imbued it with life. This is no mere symbolic representation. It is a warm and living likeness that could have well been executed at any time within the last 500 years, and is worthy in its own right to take its place with other acknowledged masterpieces.

The second example is an implement, one of a number, possibly a bodkin made from antler. I have not actually seen it, but the photograph indicates that it is marked with either 33 or 34 incised notches.

The last is a piece of wolf bone about 4" long, clearly incised with 55 notches, The series is divided at number 25 by a heavy accentuation, the lower portion comprising five groups of 5. The remaining 30 are divided into two groups, demarcated more lightly, one of 22 and one of 8.

This object has been described as a tally stick but I consider that this facile attribution is very much open to doubt.

Is it but a coincidence that the tally should finish at the Fibonacci number 55, and that we should also find the Fibonacci numbers 5 and 8 and a close approximation to 21? What are the odds that mere chance should designate the numbers? And we must also bear in mind the bodkin displaying either the approximate 34 or the actual Fibonacci number 33.

It seems very odd to me that references to numbers in a series not to be formulated for another 20,000 years and more, are so insistent. I think that we cannot treat these three coeval clues in isolation and I lay some emphasis on the first, the portrait head. It is quite unique and art at this level of excellence does not manifest itself all of a sudden. It must have evolved over a considerable length of time. Furthermore it suggests that it was produced for and amongst those who were capable of appreciating it, at a time when mammoth, reindeer and woolly rhinoceros roamed an area thinly peopled, in the popular imagination of today, with a sprinkling of uncouth hairy savages eking out a precarious existence.

(b)

The two instances that follow are concerned with transport, and are in the context of dimension.

The first is to he found in Malta where pairs of parallel ruts appear in the rocky terrain. These run for considerable distances, and assume various conformations. Various theories have been advanced regarding them, and the most convincing is that of Instructor-Captain H.S. Gracie R.N.* who published the results of his studies in 1954. He points out that whereas they were clearly used by vehicles, the great depths in some places would preclude the use of wheels unless these were of an inordinately large diameter. Sledges could not have been used because where the track was curved, neither widening nor flattening of the bottoms of the ruts, such as would be caused by sledge runners, have been found. He concludes that the vehicles in question were draw-carts, a simple device consisting essentially of two shafts supported at one end by the draught-animal, the other end trailing on the ground. These are still in use in various parts of the world.

There is a wide divergence of opinion in regard to the actual date of these tracks, but it is assumed that they are of prehistoric origin; from about 1000-600 B.C. seems to be a reasonable conjecture regarding the periods in which they were in use. The gauge varies but by far the largest number measured lie between 53" and 57", and the average is 55".

Near the village of Westhay, between Mendips and Quantocks, excavation has revealed a prehistoric sleeper road nearly 5000 years old. It consists of shaped alder logs laid transversely and secured with birch pegs. A few of these logs are up to ten feet long, but the majority range in length between 48" and 60". The average length is 54". As a matter of interest, the scantlings of these members are 8" and 5".

---

* "Antiquity" vol. 28, no. 110, pp. 91-98

(c) Stonehenge

The building of this was a major undertaking. It will be well to consider in the first place, the factors that led to its construction, or rather, reconstruction, because it is generally accepted that the work was done in more than one stage, the periods being separated by many hundreds of years.

The first thought that strikes one, after considering the magnitude of the great circle and of its component members, is that it shows evidence of prodigious "drive" or equally prodigious enthusiasm; or possibly a mixture of both.

Great numbers of people must have been involved and submitted themselves to authority; taken from the normal essential task of food production, and set to work on something that appears to us to be of little if any direct utilitarian value.

This could have been an instance of tyranny applied by the enthusiastic. On the other hand it could have been the spontaneous product of some fervent belief akin to that implied in the charming story of the building of Chartres Cathedral, when the whole countryside, rich and poor alike, turned out to help with gifts of food, material and labour. I, personally, prefer to believe in this alternative.

We shall never know the answer, but the fact remains that no other structure of comparable magnitude and shape was constructed during the next two thousand years.

The diameter of the outer stone circle is 108 feet, about the same as that of the dome of St. Paul's.

When one involves a substantial portion of a whole population in a venture of this kind it argues its own importance and it seems reasonable to suppose that this basic dimension received the most careful consideration and that there must have been a reason for the choice. Unlike the previous examples, this work is to a larger scale, so for the reasons already given it can be presumed that the basic module adopted would also be large, the "foot" instead of the "inch".

The radius, the all-important setting out dimension, is 54' 0", a close approximation to the Fibonacci number 55. If one opts for the foot of 12 reduced inches, postulated previously, it is closer still – 54/0.98 = 55.1.

The diameter is, practically speaking, 40 of Professor Thom's mega yards (39/40 group). As we have seen, these appeared to have been used in multiples of 2 ½ . So the number of mega yards comprising the radius is the Fibonacci number 8. It should be noted in passing that the verticals of the great trilithon, the focal point and summit of the design, are just about 21' 0" high, and also that the "Aubrey" holes, a ring of post holes the intent of which remains obscure, number 56.

Since the preceding lines were written, a report has appeared regarding the discovery of a Roman "arena" at Lunt near Coventry. The post holes which alone remain are arranged in circular fashion upon a diameter of 107 feet, that is a radius within an inch of 55 Roman feet.

It is unlikely that the designer (or, for that matter, Wren planning St. Paul's) visited Stonehenge in search of precedent, and then made some such remark as "*Hic dimensionam requirit est*".

Does it not seem more probable that the ruling dimension in all three cases was based upon a traditional and accepted numerical canon? This gives one to wonder how many more examples there may be still awaiting recognition.

(d) The great circle at Avebury

This lies not many miles to the North. It retains little of its Megalithic character save for a few enormous scattered stones. The earthworks alone remain to afford us what little evidence we can gather, and even this may be coincidental. The top of the rampart in the places where the surrounding ditch remains excavated to its full depth is 55.5 feet above the floor, and the latter is 40 feet across, that is, to all intents and purposes, 6 Cannes.

(e) Ancient Cosmologies

Plato was much concerned to evolve a concept of the physical universe which would accord with reason, and conform with the results of observation.

One of his contemporaries, Eudoxus, propounded a system of successive invisible concentric spherical shells revolving at different speeds around a central sphere, the Earth. These shells carried the planets, and the outermost one the fixed stars. Beyond this was the *Primum Mobile*. This theoretical mechanism seemed to work very well, but underwent some substantial modification until, in post-Aristotelian days, the number of shells had been increased and settled down for a long time as 55. We, with knowledge acquired since, can see that the whole concept was totally erroneous, albeit that at the time it was evolved by a process of extremely able and constructive reasoning. The interesting point arises: Was the end of the logical chain conditioned by a number embedded in the subconscious of those same Greeks, thus to receive the crown of acceptance and inhibit further enquiry?

This possibility deserves consideration for we find an echo some two thousand years later. Tycho Brahe, theorising upon the mutual attraction between heavenly bodies, cites that between the Earth and the Moon. By way of example he divides the distance separating them into 54 parts. This is not 55, but is very close to it.

It is interesting to reflect that the system originated by Eudoxus still retained a measure of currency right up to the end of the 16th century. Had this not been so, Shakespeare's Oberon would never have informed Puck that "Certain stars shot madly from their spheres".

(3) Eighty-Nine

So far I have discovered but one instance of this number, and at that, somewhat conjectural.

The old Jewish short cubit measures 17.72", and 5 of these give 88.6".

It will be recalled that the fathom, a special enlarged unit, contained six feet, and that the carne also contained six units equivalent to the length of the common old English building foot (which was also used much earlier in the construction of Silbury Hill) so it seems not improbable that this assembly of five short cubits could have constituted another synthetic unit of similar kind, although in its physical form it must be admitted that it would be heavy and awkward to handle – yet not impossible.

A long time ago, in the high Alps, I once came face to face with a monk, an octogenarian, resembling nothing so much as a cheerful and benevolent Old Testament prophet, attired in a long black habit, manfully if laboriously making his way up to the next refuge. He carried an enormous rucksack, and his beard was of inordinate length. Likewise his staff, because although he was a big man, it overtopped him by far. Perhaps it was with a staff such as this that Moses struck the rock.

Had it been possible to include the brazen locks that secured the gates of Tartarus, it would have been much more to the point. The number is, however, regrettably three short of requirements.

(4) 144, 233, 377

(a) One Hundred and Forty-Four

144 by itself has proved unfruitful, except for the gross, which is the square of the duodecimal base.

(b) Two Hundred and Thirty-Three

Two hut circles at Llanrog in North Wales were found to be 230 feet across. This could be taken as unconvincing insofar as the dimension concerned is the diameter and not the radius. Otherwise the correspondence is plausible enough, bearing in mind wear and tear over some thousands of years, possible inaccuracy on the part of the originators and difficulty in now defining exact measurements for an earthwork.

Nevertheless the dimension reappears elsewhere in another structure, built perhaps 2,000 years later. This is the villa at Bignor, Sussex, one of the largest in this country and famous for the surviving mosaic floors. The great court around which the various rooms are grouped is not an exact rectangle, but the West and South corridors are at right angles to one another. They measure a shade over 230' 0" and 142' 0" in length respectively and thus form the base and perpendicular of an almost perfect Golden Section triangle.

Some further examples will be found towards the end of Chapter XIII.

(c) The Great Pyramid

This monument has been the subject of so many improbable fantasies that I am reluctant to mention it.

However, it incorporates a striking example of the theory that I have formulated, and endeavour to describe.

The recovered papyri that deal with geometric problems include some which are concerned with the design of pyramids, dealing in particular with the angle of inclination, and great emphasis is laid on the primacy of half the length of the side of a square base.

A moment's reflection will reveal that this is the same thing as the radius of a circle contained within a square, touching each side at the centre. It will be recalled that the radius was the crucial factor in the setting out of Stonehenge, and that this radius is represented by the Fibonacci number 55. At Giza, the length of the base is 754'0". The half length equal to the enclosed radius is therefore represented by another Fibonacci number 754/2 = 377

We can imagine the Pyramid being contained within a rectangular solid. If the vertical sides of this are 467' 0" high, they will be Golden Section rectangles, In fact the pyramid is 479' 0" high, so that the apex rises 13 feet above the top plane of the parallelepiped.

There are various good reasons for this apparent aberration and these are given at the end of Chapter XIV.

For my part, I have no doubt that the prime concept was based on the Golden Section.

Another curious fact emerges. If we accept a dimension of 13" as representing one of the measures of the ancient foot, three of these will amount to 3.25 current feet. This closely approximates to the length of the Great U. If we divide the length of the base by this dimension we get the number 231. It only requires a minute adjustment to the length of the basic unit to transform this into 233. As may be expected, the same process applied to the hypothetical height gives the Fibonacci number 144.

(5) Six Hundred and Ten

We are nearing the upper numerical limit in the field of practical application. Akitanes may have discovered, or thought he saw 10,000 archers lurking in the olive groves, and Herodotus probably magnified the strength of Xerxes' army by ten to give the reputed total of 1,700,000, but the figures are theoretical and quoted for their dramatic quality, not their accuracy. Later writers have adapted the same approach.

Marlowe, invoking legend not fact, implies that the avenging host that set sail with Agamemnon, Achilles and Menelaus for the expedition to Troy numbered some 50,000 whereas it is far more likely that Helen's glance involved but a tenth of that number in what can be described as a somewhat rough time.

This brings to mind a delightful story regarding a visitor to the Aegean, at a chance meeting with a wrinkled salt-pickled and, incredibly ancient sea-farer discussing the fall of Troy, and backing a theory that the true causes of the conflict were commercial rather than personal.

The mariner says: "Yes, yes – what you are saying might well be true, save for one fact"

"What is that?"

And the old man, Ulysses himself replies: "You see, you are much too young ever to have seen Helen!"

To return to our Fibonacci or Golden Section Number, 610, there is indeed no evidence of its use. We can hardly expect to find it although something may yet come to light.

There is however, a possibility that it may have been utilized as the foundation for a particular dimension and as it happens, a very important one, the stadion.

The stadion or stadium is a Romanised Greek word for a measure of approximately 600 feet. All distances were measured in stadia, the ancient Mediterranean standard, just as the obsolescent mile and the intruding kilometre define distances in our own times.

The actual length is a matter of dispute.

V: Findings

One authority states "a length of 600 Greek feet, about 606 feet by our standards".

Another from an eminent source equates the total length with 624 current feet.

We thus have at least two conflicting versions, and I dare say that there are more.

The difficulty with the first is that although the "Greek foot" works out at 12.12", close enough to an authenticated measure of 12.15", this unit is of rare occurrence.

The second is more plausible, the foot concerned being 12.45" about the centre of a large group ranging between 12.40" and 12.62".

One might have expected the foot of 12.62" found at Olympia to provide the generating unit, and if this were the case a third variation of the stadion would measure nearly 630 feet.

There is thus a considerable difference between the highest and lowest estimates, about 4%, and if great distances were involved, the cumulative effect could have amounted to a very considerable figure.

We should, I think, consider practical issues in order to arrive at some reasonable conclusions. Right up to comparatively recent times, the definition of distances on the ground has been at best a matter of approximation. I expect most of us have suffered the experience of consulting a sign-post, telling us that such and such a place is three miles away, and then walking half a mile to find another one pointing in the same direction giving the same distance or something even greater.

In Ireland, as we all know, the length of the mile often appears to be inversely proportionate to the strength of the native's will to please.

In any case, how often do we check that the measure that we are given is accurate? Simple enough with commodities, but how fraught with difficulty when it comes to assessment of distances.

A marker or a mile-stone may well be out of position due to the physical unsuitability of the just site, or because, if this were selected, a lack of prominence. To a pedestrian, rider or mechanised human, 50 or a 100 yards more or less is of small import. At the true mile, the inn is but little beyond and well within reach. On such an occasion, desire for refreshment could well outweigh academic consideration of any error in actual distance.

On the shores of the Mediterranean, in the early times of which I speak, difficulty in the accurate definition of a distance would have been even more pronounced. The surveyors would have had to use rods or lines of limited lengths with all the possibilities of error implicit in this method. Waywisers had they been invented, would also have been subject to inaccuracy in the results they gave, for how could one be constructed to give true readings when the circumference was subject to but an approximate assessment of the value of π ?

Hard rock at no great distance beneath the surface covers much of the terrain and it would be natural to plant a marker in a soft patch, slightly out of position instead of going to work pecking away at a slab of basalt with soft metal tools, in the interests of strict accuracy.

It is therefore this context and possibility of error that makes nonsense of any modern evaluations of the intentions of those who originated the stadion; such assessments must of necessity be conjectural, so there is thus no reason why I should not propose yet another – that the stadion comprised 610 basic units. The length of each of these could lie anywhere between 11+" and 13+"

The spectrum of variation being in the order of 33 yards, the median for the total length would be about 625 current feet.

(6) RANDOM EXAMPLES

(case a) Vaison-la-Romaine in Provence possesses a small and strange cathedral church. This building is an example of the naive application of dimly understood principles, and falls little short of being an architectural calamity. The earliest part is the apsidal east end built at the beginning of the early 7th Century: The remainder is some six hundred years later in date.

V: Findings

This apse is a dreadful business in more ways than one. It is founded on a loosely piled heap of fragments from earlier building, roughly cemented together.

Classical architecture does not appeal to me. I can admire it but I cannot say that I greatly love it, yet I found the sight of all these fragments of richly carved capitals fluted columns and decorated frieze degraded to such an ignominious usage shocking in the extreme.

On this foundation, the lately-heathen, afire with new belief, proceeded to build the wall, semi-circular in plan as closely as they could get it and punctuated internally with a group of five rounded arches. This posed the problem of building arches curved both horizontally and vertically. Whereas but a few hundred years earlier the masons of Imperial Rome would have taken it in their stride, the Gauls found themselves in difficulties. The result is about as uncouth as can be imagined.

At my last visit, I found the sermon dull, and attention wandered to the contemplation of those five arches. Two were embellished with later accretions but the others were as their builders had left them.

The first comprised 22 voussoirs, and the other two were built with 21.

This prompted me to examine the chancel arch, a grim uncompromising pointed affair of the late 12th Century. I well remember the mental jolt, followed by a feeling of excitement, when I found that this was composed of 54 stones, 27 a side. By the time the masons had reached the top there was no room left for a key-stone, otherwise there would have been 55.

Of late I have extended this kind of examination to the chancel and tower arches of those small parish churches near my home that retain much late 11th, 12th and early 13th century work. In every case, I have found that the Fibonacci numbers from 3 to 21 were employed, either alone or in combination with one another. In two instances the 39/40 group appeared as well. It would seem that for some as yet unexplained reason, this selection of the number of stones to form an arch had some symbolic significance.

To emphasise how such a tradition can persist, either at or below the level of consciousness, I refer to the tower of St. Martins-in-the-Fields, Trafalgar Square. This was built in 1726 by Gibbs. The arched openings to the tower comprise 21 projecting quoins and voussoirs, the number including the triple key-stone.

(case b) Some 40 km. away from Vaison lies the small and attractive town of Saint-Rémy where Vincent Van Gogh was under medical care towards the end of his life. To the south are those charming miniature mountains, Les Alpilles, sheltering the ruins of Glanum, a site first occupied by aboriginals, then Greek settlers, and later the Romans. The ruins have yielded a rich harvest of coins, jewellery, pottery and domestic articles, much of which is housed in the museum at Saint-Rémy.

One of the objects that caught my fancy was an ivory or bone dice cube some 2500 years old. It was exactly the same in appearance as a modern one, and perhaps it was this evidence of continuity that prompted my attention.

I thought of Palamides at the siege of Troy, of Varus who cast dice for days and nights in succession, and of Caligula, who apart from other whims and excesses, including the bestowal of the consulship on his horse, Incitatus, is said to have been a notorious cheat at the game. But it also showed that both ancient and modern dice have a common characteristic in that the number of indices on the six faces add up: $1 + 2 + 3 + 4 + 5 + 6 = 21$

The six faces of a dice cube adumbrate some sort of an analogy with the measuring devices mentioned previously, two of which comprise an aggregation of six basic units. Here we have an aggregation of 6 groups of numbers making up 21. These groups are not merely for the purpose of identity, like those of house numbers in a street. They represent differing values. The "six" exercises three times the influence of the "two".

This number 6 is intrusive. It occurs in other contexts as well, notably that of coinage and capacity measure (see Chapter VI p.45) It defines the characteristics of any parallelepiped in regard to the number of enclosing planes, and the earliest geometers must have been both pleased and satisfied when they found that the radius of a circle could be used to divide the circumference into six exact segments.

(case c) In the Prado museum there is a prehistoric sculpture in the form of a short limestone cylinder.

Near the top are two radial designs representing eyes perhaps, or breasts. Each of these is composed of 13 strokes.

(case d) Reverting to mensural characteristics, the small twin obelisks in front of some of the Egyptian tombs measure 54½" and then jumping forward some thousands of years we find a numerical instance in that the shares in 16th Century trading ventures were subscribed in the odd proportion of fifths, eighths, and thirteenths (see also Chapter XI).

(case e) Capacity measures have so far yielded little beyond the first example mentioned in Chapter I ...

However, I mention the following:

The Syrian Kotyle has two values, 20.8 and 21.5 cubic inches, and the Babylonian and the 18th Dynasty Egyptian log appears to measure 33.2 cubic inches. The old wine gallon measures 231.0 cubic inches.

And here are a few more gleanings in the context of linear measurement.

The "Pied du Roi"	12.75 inches
The "French"' foot (Canada)	12.80 inches
The Vara (S. & Central America)	32.99 - 33.33 inches
The Madras "Guz"	33 inches
The Ken (Thailand)	40 inches

39/40

I have referred several times to a measure of 40" or thereabouts. It is very old, so old indeed that it had long fallen out of use before being re-invented and adopted by the French National Assembly as the metre, in 1791.

This number does not appear as a Fibonacci term, but is evidently related to it (*see:* Chapter IX) and seems to play some part in the Design of the Great Pyramid. As I have said it is a nonconformist amongst the orthodox, the odd man out.

When the Fibonacci series is extended up to the 107th term, it repeats itself with fair accuracy en route. This matter will be gone into more fully in Chapter VIII but it is of interest that within the range, the 39/40 figure occurs, and repeats itself three times. The odd thing is that the median occurs near the top of the range where it is expressed as: 39.28…

Our methods of measuring the elapse of time depend upon the rotation of the Earth once in a period that we divide into 24 hours. Of late it has been found that this rotation is not entirely regular, but it remains accurate enough to serve our practical needs. The period cannot be related to our journey round the Sun because this does not occupy an integral number of days. Each time we go round we lose about six hours, hence the necessity for leap-years.

Nevertheless, we can accept the rotation of the Earth as part of the natural order; but by no manner of means our sub-divisions of the time occupied, into hours, minutes and seconds.

The second has no counterpart in nature and owns its existence to the sexagesimal systems (that based on the number 60), devised in Babylonia so long ago.

From this we also derive a still prevalent system of angular measurement involving 360 degrees, each degree comprising 60 minutes and 60^2 seconds.

From the 13th Century until recently, our measuring instruments, mechanical clocks, have depended upon a mechanism comprising a pendulum and an escapement. The pendulum regulates the intermittent movement of the escapement and the following train, and the escapement returns the compliment by imparting those quanta of energy necessary to keep the pendulum swinging at a regular rate. The pendulum invokes the influence of gravity, an inexorable elemental, derived from the rotating mass of the Earth, and the gravitational acceleration (32 feet per second per second in vacuo) consequent upon this. The rate at which the pendulum swings is determined solely by its length, and it is found that to induce a rate of one swing per second, the length must be 39.14", no more, no less.

Now, a curious coincidence follows. The length of the pendulum corresponds very closely with the length of the Babylonian Great U. It is also little short of the current metre, a dimension which was arrived at independently, some thousands of years later. Incidentally, this 39" dimension divided by the Golden Section equivalent of Fibonacci number 2 (1.88) gives us, within one-eighth of an inch, the length of the Royal Cubit.

There is much food for thought in all this, because the culture originally responsible for the numerical system upon which the concept of the second is based, also made use of the Great U as a unit of linear measurement.

It would be pleasant to deduce that the ancient Magi experimented with pendulums, and adopted the optimum length to serve as a standard dimensional unit.

Unfortunately this attractive theory has to lie rejected on the grounds that pendulums as time-measuring instruments were only put to use some thousands of years later, and, in any case, it is hardly possible that that fine division, the second, was so defined in this remote era.

It is equally absurd to assume that the ancient Babylonians in some manner or another equated the period of the Earth's diurnal rotation with that taken by an actuated pendulum of a certain length swinging 86490 times.

There is another odd fact. Johan Bode in the 18th Century propounded a relationship between the distances of the planets from the Sun, measured in "astronomical units", the Earth's distance having the value 1 AU

This, Bode's Law, involved some arithmetical juggling. It has been found that the Fibonacci numbers represent the distances better, although somewhat roughly:

e.g. (*inter alia*)

Earth	1.0
Jupiter	5.2
Saturn	9.5
Uranus	19.2
Neptune	30.1

Pluto, somewhat to everyone's annoyance, refuses to fit into the sequence at 55, or anything near it. He, the god of both Death and Riches can afford to be a nonconformist, and pursues his circumsolar course at a mean distance of 39.5 AU.

This is an attractive concept, but regrettably untenable, for Pluto's orbit displays a measure of eccentricity unmatched by that of any other planet. Indeed it is so exorbitant that on occasion he crosses and recrosses the orbit of his nearest neighbour, Neptune.

CHAPTER VI

AND SOME COMMENTS

(1) Bases

We count in tens, a decimal arrangement in use since the beginnings of History. Counting by the score is no doubt a variant, but counting in dozens is not, although this in its turn could be related to the excellent ancient Babylonian system of counting in multiples of 60.

This last can be regarded as an example of inspired thinking. At a time when fractional divisions, as we know them, were difficult of comprehension, those ancients hit on the idea of choosing a round number, sixty, that could be divided in many different ways. The system is still used by us in the measurement of time, and angular displacement. The bull-wheel on a modern lathe has 60 teeth to allow a wide range of settings.

We should thus expect to find that the measuring systems in use now, and derived from the past, would confine themselves to ranges of numbers deriving from one, or other, or both of the systems mentioned above, and exclude any number that cannot in some way be directly related to 10 or 12.

We find, however, that this is not so. We find that non-compatible numbers, including the Fibonacci numbers appear in all kinds of unexpected places.

This could be a matter of coincidence, but when, as I have shown, these numbers appear in many different contexts, it would seem that this must be attributed to something more than the working of chance.

(2) Accuracy of measuring units

It can be argued that many of the examples do not exercise strict compliance. As I shall show later, there is a certain ambiguity in the possible choice of a number, but in addition to this we must consider not only the technology of early times, but the changes, unrecorded, that could have come about since.

We find measures – the 12/13" foot is an example – used in Britain, Ancient Greece and Hither Asia, varying somewhat in length, but all bearing a strong family relationship.

The propagation of a standard unit through poor communications over great distances, could account for the differences. The physical measuring devices involved, the wooden staff or knotted thong would be subject to changes in length, by reason of humidity, temperature, wear and tear or just plain misuse, and of course the possibilities of variation when reproductions were made. An incorrect version could be accepted, absorbed into the vernacular, and become a standard in its own right.

Again, although our remote predecessors were imbued with the idea of precision, we cannot imagine that this applied everywhere and at all times.

Even in these days, when the ideal of strict accuracy is so much sought after, we do not always get it.

I have three "engine divided" draughtsmen's scales. These are precision instruments, yet they do not strictly compare. The difference between their reputed readings is, admittedly, very small, but it can be detected by the naked eye.

A recent digest prepared by the Building Research Station is devoted to the subject of accuracy in setting out, and examining the pitfalls that still exist even with the use of the most up-to-date equipment.

Some few years ago I had a meeting with a builder to peg out the profiles for a new house. I brought my new steel tape and he his much worn linen one. When we compared them it was found that his had stretched an inch or two in fifty feet, yet, perhaps suspecting some hidden disadvantage in using mine, he opted for the use of his own, with the result that I got several hundred extra cubic feet of building, I think, at no additional charge.

If we could go back some thousands of years I am sure that we would find that measuring units, all bearing the same name, would differ in length from place to place, even those within relatively short distances of one another. That the ancient measuring units, so closely alike, should differ somewhat in length is in no way remarkable. Considering the circumstances, it is a wonder that they do not differ more. It seems to suggest that there was an accepted norm; that the discrepancies we find were due to accident, rather than intent.

(3) Choice of Numbers

The number of days allotted to the year is out of phase with the time taken by the Earth on its annual journey. Nevertheless, we have no choice in the matter. We have to accept the daily round and accept the nearest practical compromise – 365 days for 3 successive years, 366 for the fourth, and a further adjustment every 400 years. The number of days in the week is of a somewhat different order. Here we have chosen the Moon as a clock in place of the Sun, but we could have chosen otherwise – a year comprising 17 months of four five-day weeks plus one month containing 5 weeks.

Where freedom of choice is not restricted by external considerations there appears to have been a tendency to concentrate on particular areas in the lower levels of the numerical sequence.

Almost all whole numbers between 1 and 13 have been invoked in this manner. The 3, 4, 5 group is one, 7, 8, 9 another, and 12, 13 a third, with particular preferences for integers 3,5, 7, 10, 12 and 13.

Above this level, we find a hiatus until we come to the 20,21 group, another between this and the group centred in 33/34 and then a gap between this and the 39-40 partnership. Higher still, the lacunae increase in size, and save for some numbers stemming from decimals or sexagesimal bases, and the Fibonacci numbers, there appear to be very few preferences.

Naturally there are some exceptions. Number six appears to be popular within a limited context. As we have seen it appears to have been greatly favoured when it came to aggregating units to form what can be called portative measuring devices. It also appears in sundry ancient monetary (weight) and capacity sequences, *viz*:

6 obolos	1 drachma	(Hellenic)
6 Kats	1 saton	(Phoenician)
6 Kyathos	1 Kotyle	(Attic)
6 Siliqua	1 scripulum	(Roman)

As we are talking about "six" this is the time perhaps to mention the long-bow, which, so the experts tell us bore little resemblance to the elegant confections of today. Tradition gives its length as six feet, but is silent in regard to the length of the foot involved. This weapon which decided the outcome of so many battles, was rough, knobbly and crudely fashioned, valued for its sinew, not its looks.

We can tabulate numbers that have served as a staple for choice, have served utilitarian purposes, and are also quoted in tradition and legend. Here they are:

3, 4, 5, 6, 7, 8, 10, 12, 13, 20, 21, 30, 33,
34, 39/40, 50, 55, 60, 100, 144

When we delete those that have decimal or sexagesimal associations we are left with the following eight, of which six appear in the Fibonacci series:

3, 4,

7, 8,

13, 21, 33/34, 55

We should add back 5 and 144, and in fairness, I think, the related 39/40. Thus we have

3, 4, 5, 7, 8, 13, 21, 33/34, 39/40, 55, 144

So, out of 11 numbers, eight belong to the series, and the ninth is related to it.

The two remaining have always received their meed of homage and need little comment. 4 is a kind of numerical jack-of-all-trades and 7, the 'lucky' number, is a kind of incantation in itself.

(4) Leonardo of Pisa (Fibonacci) circ. 1170-1230

We are confronted with a paradox. This great mathematician is credited with the formulation of the series in the 13th Century, but as I have indicated, there is ample evidence to prove that its unique character was known many thousands, and possibly tens of thousands, of years before his birth.

I advance the following alternative suppositions:-

(a) Fibonacci re-invented the series. This sort of thing is not uncommon. The value of π is commonly supposed to be an Hellenic discovery possibly because of the symbol which distinguishes it. But both ancient Egyptians and Babylonians had arrived at practically the same conclusions at least 2000 years beforehand. We hear of identical discoveries being made quite independently. I myself had a personal experience of this sort. Once when conducting a search relative to a patent of my own, I found that an American inventor had arrived at the same conclusions 14 years previously. Not only was the form of the device the same, but the actual dimensions and angular characteristics as well.

(b) Fibonacci may have been led to his formulation by the same kind of observations that form the substance of this book, and in consequence codified the series in a way akin to Euclid's arrival at the definition of the Golden Section. He may well have been assisted in this by source material since lost. This is not at all unlikely. Such, before the advent of printing, would have been a rarity in manuscript form, and there would have been a high probability of loss or destruction during the ensuing troublous 700 years.

(c) The most attractive theory is that of an hereditary aptitude and the working of race-memory. The genes which, so we are told, transmit memory traces and characteristics, move in deep and tortuous paths. Sometimes they pop up after a generation or two, at others, they may efface themselves. Perhaps, who knows, they may delay their overt influence for centuries or even millennia.

This conjecture admits of no possibility of scientific proof insofar as few people can trace their ancestry back far enough in a reliable fashion, and fewer still are gifted as prophets. Nevertheless, I can point to a slender clue.

I refer to a Romano-Grecian copy of a bust still earlier in date, purporting to be a likeness of Pythagoras. From early times, apart from the mythical characters such as Orpheus, Apollo and Marsyas, the name of Pythagoras alone is that remembered in association with the art of music.

Others there are, but they receive scant mention. For all we know they may have possessed talent or even genius, but no hint remains of their achievements. This is all very strange. It is as if, in the 45th century, the name of Helmholtz should be remembered and honoured, but those of Beethoven and, say, Josquin forgotten.

However, Pythagoras must have combined musical and mathematical insight or he would not have made the discoveries that he did, and lay the foundations of the science of acoustics so long ago.

I now come to compare the bust with a portrait painted about 1905 of Claude Debussy, that outstanding musical genius and innovator, who changed the whole course of the art, and by so doing regrettably inaugurated the process of disintegration that now prevails.

The bust and the portrait display an uncanny facial resemblance, and I am led to the surmise that were it possible to trace Debussy's ancestral line back for some 100 generations, it would lead direct to Pythagoras.

If there be anything in this, we could in like fashion conjecture that the long forgotten author of the Golden Section and its related numbers might have been an ancestor of the great mediaeval mathematician.

Thus, after countless centuries, the mind of Fibonacci could have constituted the ready and receptive mirror in which the ancient concept was reflected.

As a postscript to this section I should mention that Leonardo's book "*Liber Abaci*" introduced "Arabic" numeration into Europe. He is thus in part responsible for nearly all the material benefits we enjoy, and also the current ills with which we are afflicted.

CHAPTER VII

INTEGERS & FRACTIONS

It will have been noticed that in many of the quoted instances the dimensional equivalents of the Fibonacci numbers depart from the norm. Sometimes the difference is very small, but in other cases they can amount to 2 or 3%.

I have already referred to inaccuracies that could be brought about by the process of time, and other causes, but an additional factor has to be taken into account.

We are accustomed to the use of fractions, in vulgar or decimal form, although it should be realised that even now, many intelligent and otherwise well-educated people, mainly the elderly, or those of middle age, cannot cope adequately with decimal fractions.

We are also forced into the position of regarding former ages from the standpoint of our own, and it is sometimes difficult to realise that what we take for granted would, in many cases, be unknown or unintelligible to our ancestors.

Not many years ago the significance of an expression such as "1.414" would have been incomprehensible to the majority, both in form and in meaning. At any time before, say, 1300 A.D., it would have been utterly and entirely meaningless to anyone in Europe, even the intellectual elite.

On the other hand, the use of vulgar fractions has a very long history, but in early days the knowledge was in the hands of the few, who were concerned with the conservation of tradition, not with the education of the general multitude. In any case, it is doubtful whether the latter would have possessed the mental equipment by means of which they could derive benefit.

I do not say that the simpler fractions, 1/2, 1/4, 1/3 etc. were enclosed mysteries. Indeed the Romans displayed a loveable inconsistency, and, counting decimally, divided unity into twelfths. But, it seems to me that the tendency would be to avoid fractions and rely on the nearest whole number to provide accuracy enough.

In earlier times still, commodities would be bartered in batches of whole numbers – livestock for instance. No one in his senses would demand 30 ½ living sheep in exchange for a daughter. He would either settle for 30, drive a hard bargain to obtain 31, or perhaps accept something else to make up the difference – a skin of wine or a new finely-polished axe-head.

Distribution of food would follow a similar pattern. Meats would be apportioned according to rank, or the need of the recipient. Liquid nourishment in some sort of conformity with the size of the receptacle. None of these sub-divisions would be made with meticulous mathematical precision.

A herdsman marshalling his herd of 40 goats would not recognise the concept that one goat was 1/40 of the herd.

This inability to deal with small divisions, amounting almost to resentment, has persisted into the present century. A few years ago I asked a skilled workman to mount a ladder and measure the distance of a pipe below ceiling level.

> "What is it?"
>
> "One foot five and a bit"
>
> "Can't you give it more accurately?"
>
> "More akkritly? Orl right then. (pause) One foot five, and free of them little bastards".

Precision to the verge of the microscopia is a modern phenomenon. Without it most of the sophisticated if ephemeral equipment upon which we place so much value would be lacking. It is entirely dependent upon the evolution of a decimal fraction, content for the practical implementation of ideas.

VII: Integers and Fractions

The ancient Greeks and the Mediaeval builders were highly intelligent people yet the machines that they possessed were of the crudest description, and even as late as the 16th Century, the inventions of Leonardo da Vinci were in the main prophetic, and not put to practical use.

I am certain that all this stemmed from the fact that sophisticated manipulation of algebraic and numerical symbols together with the consequent ability for fine division, although once perhaps the property of eclectic circles, had been lost sight of and fallen into disuse. On the other hand, the ancient peoples may have been even wiser than reputation admits.

It should be noted that the Hellenes, their immediate successors, and later the Mediaevals, were expert geometers. The degree of precision required for their needs could be obtained by geometric means, with little cause to call upon the mathematical symbol for aid.

All that is needed to construct a right angle is a compass. Together with the set-square obtained therefrom, and two straight edges, a given length can be divided into any required number of precisely equal divisions. All manner of curves can be set out by purely graphic means, with the aid of these simple tools.

Where numeration was invoked, this would of necessity be simple also, and irrationals resentfully accepted just because they were inescapable.

To drive the point home, it can be recounted that Pythagoras and his followers, dissatisfied with the explanations of natural and supernatural phenomena current in their time, appealed to the concept of number and attempted a new quasi-mystical synthesis.

They worked perforce with integers, that is, whole numbers. This went very well until the inevitable snag appeared. A square can be constructed, each side comprising the same number of whole units. When the length of the diagonal is considered, no assembly of these whole units will fit exactly. There will always be a bit over.

Legend states that this shocking discrepancy was hushed up, and remained a closely guarded secret amongst the elect in an effort to preserve a beautiful if untenable viewpoint from which to regard the physical and spiritual world.

The Greeks must have been fully aware that the relationships between the diameter of a circle and its circumference was not precisely $3\,^1/_7$. But because a method of arriving at the more accurate 3.14159... etc. was not available to them, they represented this irrational number by the symbol "π".

In similar fashion, Fibonacci could have expressed this series as 3.00, 4.85, 7.82 and so on. This would have been more accurate but despite the fact that the decimal fraction had already made an appearance, he conformed with the convention of his time, and adopted integers.

Such a course of action by the foremost mathematician of the day seems to suggest that before this, expert and layman alike would tend to eschew the fraction in favour of the round number.

CHAPTER VIII

PARAMATHEMATICS

This chapter is not as formidable as its title might suggest. Even readers who are temperamentally averse to enquiry of a mathematical nature should read it now, rather than skipping it altogether or coming back later. It presents few barriers to comprehension and has a considerable bearing on the whole of the subject matter treated subsequently.

Before we dive in, there is, however, one comment that must be made. In the following pages, the terms "**Golden Section number**" and "**Fibonacci number**", together with the individual numbers themselves, may appear to be used somewhat indiscriminately. There is nevertheless a practical reason for this, not just a matter of convenience (although this comes into it), but an effort to maintain clarity of meaning. Whereas the Fibonacci number comprises a whole integer, the Golden Section number does not. Each G.S. number involves the use of a whole number and a long decimal tail, a configuration much less easy for the reader, or anyone else, to comprehend.

Example: F. No.	3
G.S. equivalent (one of many)	3.06504504

Golden Section serial numbers can emanate from any of the innumerable starting points in the decimal numerical scale. We can select any number we like, for example 2.6748 or plain 973, and, by multiplying by 1.618... commence a series. The reason that even these numbers would not be entirely correct is due to an inherent flaw in the decimal context which forever precludes the attainment of an unambiguous datum. This matter will be touched upon later.

To put not too fine a point on it, this book is mainly concerned with a particular series that, closely associated with the so-called Fibonacci numbers, evolved long before the fine edge of the blade of sophisticated decimal calculation had been ground and sharpened.

(1)

This part of the investigation must start with a determination of the position of the Golden Section. Two methods are available:

a) A tedious and exasperating process of trial and error, involving much measurement and re-measurement, with no guarantee of accuracy at the end.

b) By means of calculation, so:

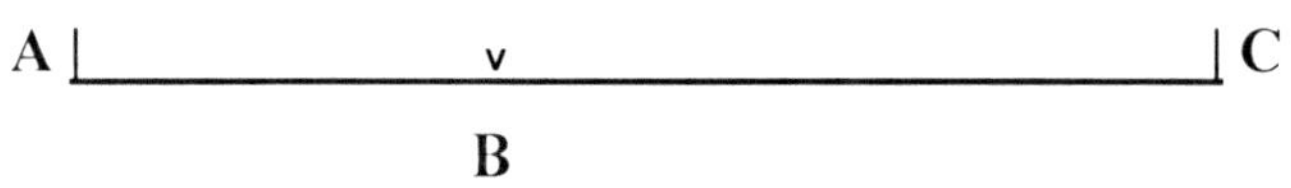

The length of A-C is to be cut at B, so that AB: BC as BC:AC

Let $$\mathbf{BC/AB = X}$$

$$\mathbf{X^2 - X - 1 = 0}$$

This equation resolves itself in the form:

$$\mathbf{X = \sqrt{1.25} \pm 0.5}$$

Either of the two values thus obtained (round figures 1.62 and 0.62) can be used. For instance, if we divide the Fibonacci ordinal 8 by 1.62 we obtain 4.96, and we get the same result if we multiply it by 0.62. In each case, the residue is 3.04.

Whatever the length that we treat in this fashion, it will be divided for all practical purposes into the proportions that define the Golden Section.

However, strictly speaking, this division can never be made with absolute precision because of the difficulties attendant upon the extraction of $\sqrt{\mathbf{1.25}}$.

If you were to instruct me to set about this and came back for results ten years hence, assuming that I had survived the onslaughts of time and writer's cramp, you would find the task unfinished.

If I had been rich enough to hire a computer for the job, to produce 10^{5000} times the amount of calculation, the ultimate goal would be no nearer. It is unavailing to blame decimal fractions for this. Vulgar fractions would be even more intractable.

The square root of 1.25 is one of the family of irrational numbers. It can never reveal its full identity. There are others of the same kind, π for example; but whereas an investigation to reveal the absolute value of π commences with two factual premises, a diameter and a related circumference, the Golden Section is based on no physical data whatever.

You can inscribe a circle and its diameter and visualise the problem. But when you draw a line, whatever its length, its division into the required Golden Section proportions rests on an intangible basis.

We can ruefully state that the more we work on the investigation of an irrational number, the less meaningful it becomes. The expression 1.62, or even more accurately 1.61803, can be reasonably comprehended. An extension of this expression to the 50th decimal place, as could be accommodated in the width of this page would become meaningless, yet this would be but the commencement of a beginning.

If we assume "infinity" to be a "fixed" point, we can indeed say that our process of calculation is recessive, and we, the initiators, move further away from the goal with every step we seek to take towards it.

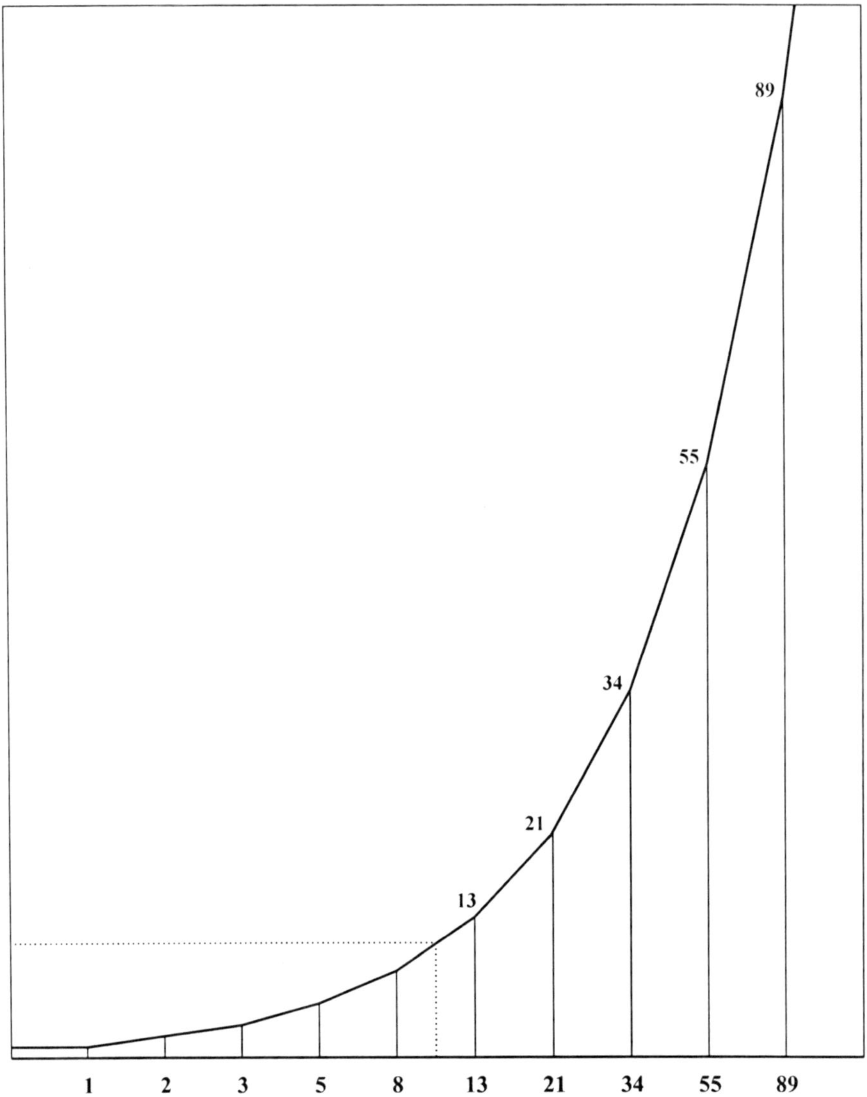

Figure 6

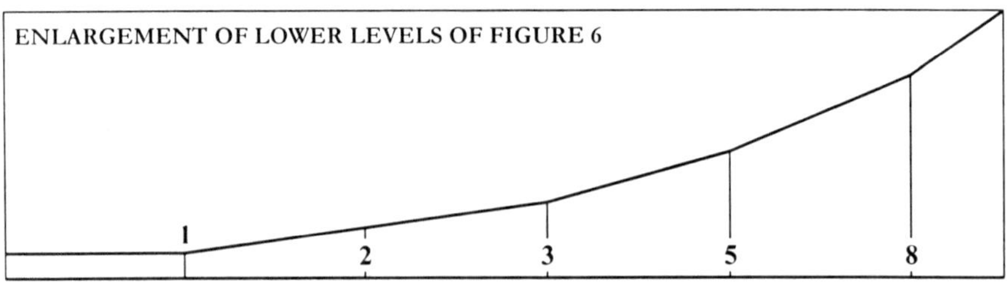

Figure 7

(2)

Our series stands upon the basis of an irrational number, so it follows that it can never complete itself. We can go on building it up endlessly.

Figure 6 shows the commencement of such an operation.

The Fibonacci numbers are represented to scale by the height of the ordinates, and it will be seen that commencing at 5 and working to the right, the resultant plot conforms with a smooth curve which gets steeper and steeper. It is asymptotic and should it ever achieve perpendicularity, the ordinate in that position would be of infinite length.

The line joining the points has no significance except to emphasise the characteristic of curvature.

To the left of 5, commencing at point 3, and working to the left as far as 1 there is a somewhat violent change in the character of the curve, and at point 1, this curvilinear character disappears entirely.

Figure 7, an enlargement of the portion under consideration, clearly shows this.

The implications will be dealt with in Section 4, but here I would draw attention to a resemblance between what the diagram indicates and a certain subjective phenomenon in the realm of motion, one which is a common enough experience. If one is driving over a long straight road, and an approaching vehicle comes into sight, its apparent movement is at first barely perceptible. After a little while it can be seen to be moving and appears to grow larger at a steadily increasing pace. There comes an instant when a sudden and violent apparent increase in size and speed takes place, immediately preceding the moment that it passes.

This illusory impression is echoed by the descending curve, orderly down as far as point 3. This is the "change of state" point, where smooth approach seems to change during a split second to sudden onrush. Point 1 is the moment of cessation.

I for one have never risked my life by turning right round at this point to see if a reverse process holds good.

(3)

We must now leave our examination of the upper register for the time being, and turn our attention to what happens at the lower end.

Figure 7 shows the nature of the change at point 3.

Below this point the only integers that remain are 1 and 2, so if we are to adhere to Fibonacci's mode, we have no alternative but to make use of them.

The result is the plot line between 3 and 1.

When we propound the statement "21:34 as 34:55" we are not being strictly truthful, but we are not far out, the error being of the order of 0.12%. When we say "3:5 as 5:8", we have crossed the frontier between truth and falsehood, because the factor of error has risen to 4%. "2:3 as 3:5" is worse; it comprises an inaccuracy of 10%, and when we state "1:2 as 2:3", the element of falsehood cries aloud at 25%. No comment is required on the final statement "0:1 as 1:2".

We could perhaps accept this distortion of the truth as temporary, and proceed with our plotting below the datum line, making use of negative integers, in which case the diagram would assume the configuration shown in Figure 8.

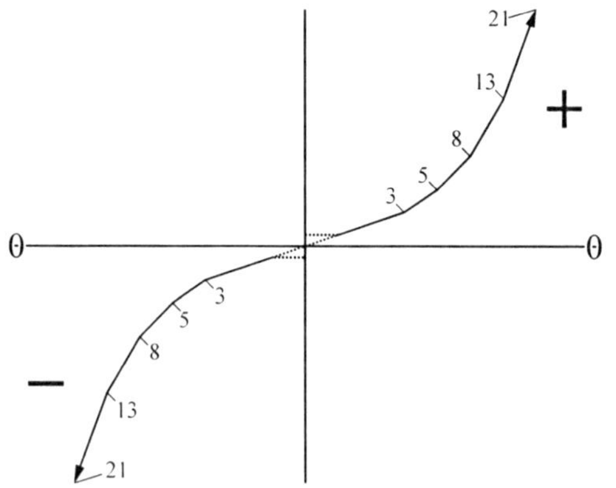

Figure 8

This might be the answer if we were dealing with things represented by imaginary numbers, ghosts seen in a mirror, but we are not. We are considering "real" things, "real" rectangles, the sides of each of which are represented by "real" numbers, as shown below in Figure 9.

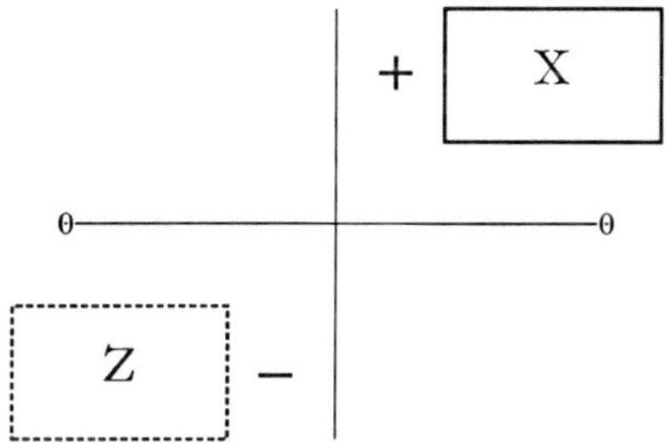

"X" IS THE "REAL", POSITIVE RECTANGLE. Z IS THE NEGATIVE ONE; IN A MANNER OF SPEAKING, X'S OPPOSITE NUMBER.

Figure 9

If we were to assume 0 as a vertical datum instead of an horizontal one, we could make use of the mirror simile and indicate this state of affairs as is shown below in Figure 10, and plot the curves as shown in Figure 11.

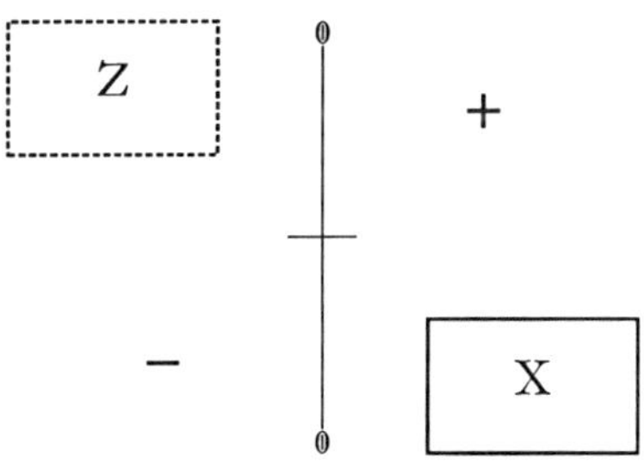

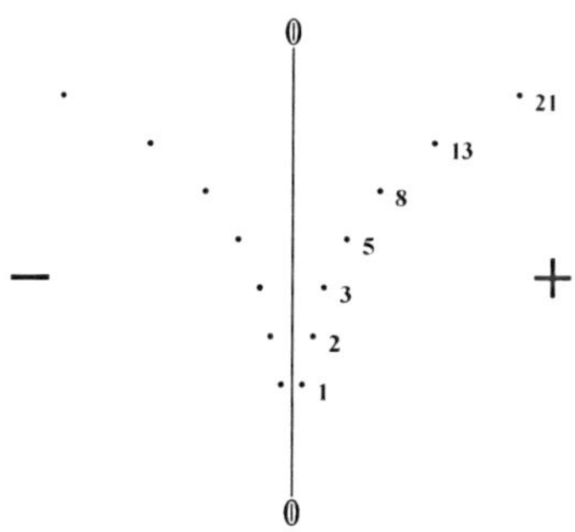

Figure 10 **Figure 11**

However, not one of these diagrams is satisfactory because the intervals between 1 and 0 and 0 and -1 cannot be plotted to scale. Such a course would require paper of infinite size.

(4)

There remains but one way out of this difficulty, that is, to abandon the Fibonacci numbers altogether except for a single initiating integer. Even this method is not sure because the infinite nature of the series precludes the possibility of choosing a definite starting point. However, if we commence at a reasonably high figure we shall achieve quite reasonable results, even if they do not comply with the demands of the absolute.

If we divide any Fibonacci number by 1.61803, the dividend is very close to the numerical value of the next lowest number. This divided in the same way, gives the number below, and so forth.

Starting at 55, the equivalent of 3 works out at 3.065. I do not propose wearing you with the details, but you can check the calculations yourself if you so wish:

F. Number	Golden Section Equivalent
55	55.00
34	33.991
21	21.008
13	12.983
8	8.024
5	4.959
3	3.065

Below this level we get

$$\frac{3.065}{1.61803} = 1.894$$

$$\frac{1.894}{1.61803} = 1.1707$$

$$\frac{1.1707}{1.61803} = 0.7235$$

$$\frac{0.7235}{1.61803} = 0.447$$

This series of calculations can go on indefinitely.

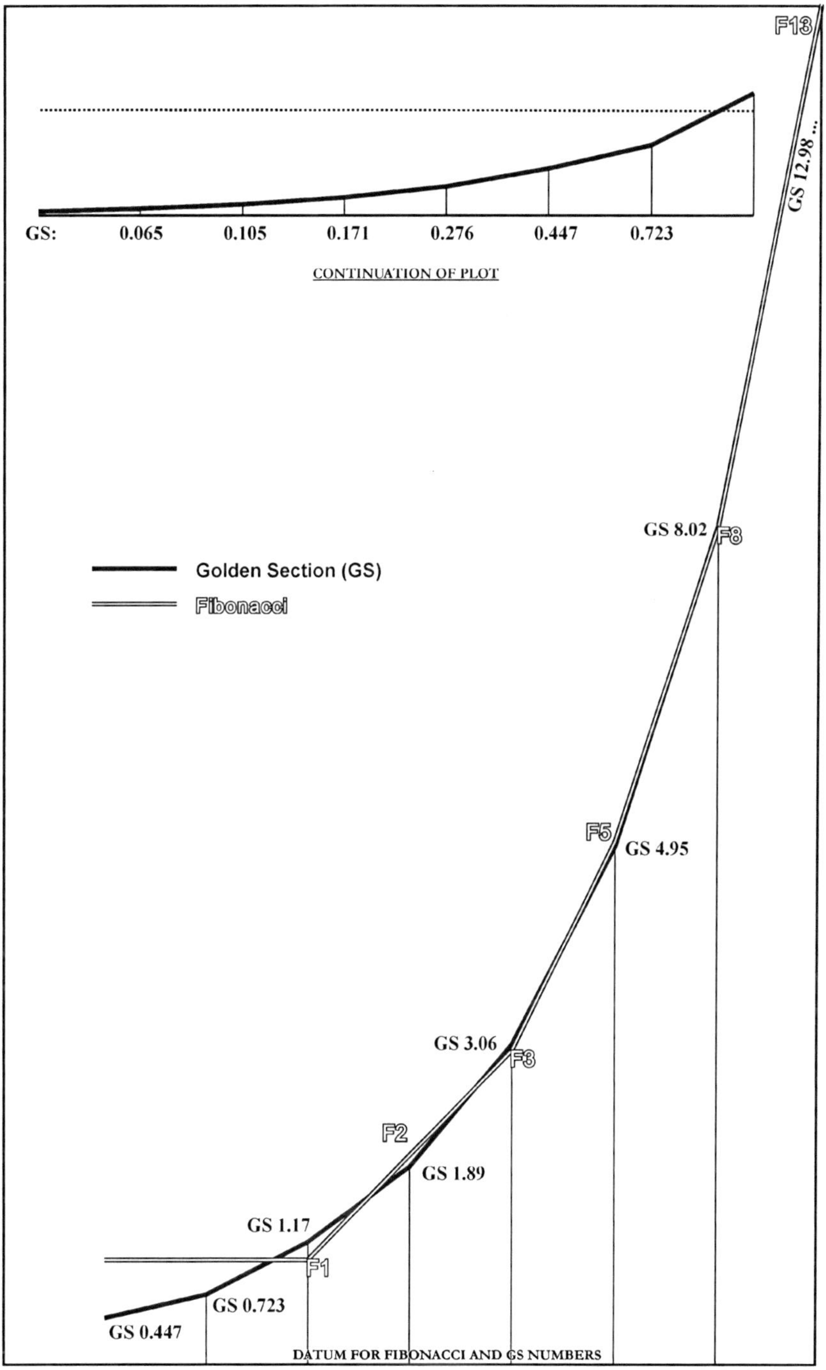
F13
GS 12.98 ...
GS:
0.065
0.105
0.171
0.276
0.447
0.723
CONTINUATION OF PLOT
GS 8.02
F8
Golden Section (GS)
Fibonacci
F5
GS 4.95
GS 3.06
F3
F2
GS 1.89
GS 1.17
F1
GS 0.723
GS 0.447
DATUM FOR FIBONACCI AND GS NUMBERS

Figure 12

In order to show the two modes, I have had to draw another diagram, Figure 12.

This indicates the erratic behaviour of the lower Fibonacci integers by a pecked line, and the suave conduct of the Golden Section numbers by a full one. In this case it will be observed that instead of an abrupt change at point 3, the curvature is now continuous.

This time the plot is asymptotic in relation to the base. It approaches zero ever more closely but can never attain to it.

Before we consider the implications of this, it will be as well to discuss another aspect of the matter, the context of error within the Fibonacci series itself.

I propose to show that this follows a rather curious pattern, in so far as that the successive relative differences exhibit what can be termed positive and negative characteristics.

It can be likened to a man shooting at a mark, the first shot going too high, the second too low. The third is a high one again, but closer, the fourth low, but closer still.

This has been plotted out in diagram Figure 13a, and it will be understood that in order to obtain values by which results can be plotted, it is necessary to make use of expressions, thus:

3:5 as 5:8

In this case, cross-multiplying 8 × 3 = 24 and 5 × 5 = 25, the error is therefore the difference between the two, 1/25 or 4% and the expression centres on the Fibonacci ordinal 5. The table below the diagram gives the numbers, the pairs of expressions and the degree of error from F2 upwards.

It will be appreciated that it is not possible to plot 1, because the expression centring on this number includes 0.

The factor of error is infinitely great and defies any method of indicating it. Indeed, the trace that the plotting below 3 indicates has to be regarded with suspicion. Is it concerned with figures or figments?

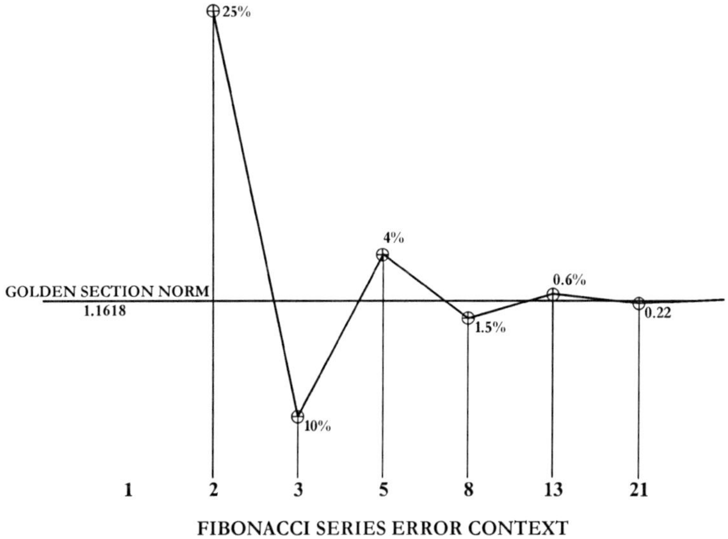

Figure 13a

F. NUMBER	RATIO	% ERROR
-	0:1 as 1:2	INFINITY
2	1:2 as 2:3	25
3	2:3 as 3:5	10
5	3:5 as 5:8	4
8	5:8 as 8:13	1.5
13	8:13 as 13:21	0.6
21	13:21 as 21:34	0.23
34	21:34 as 34:55	0.11

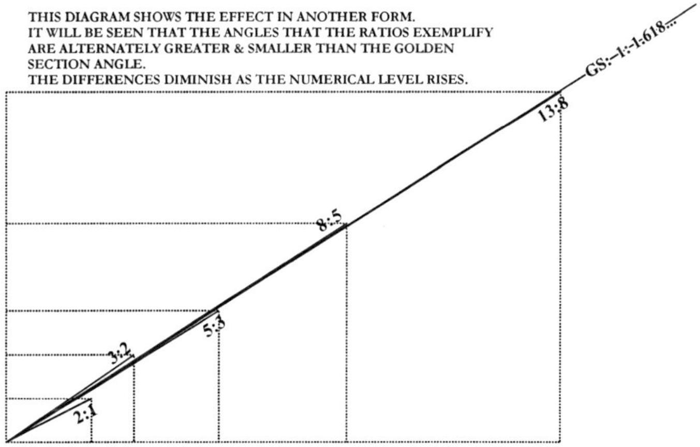

Figure 13b

The diagram does however clearly show the shortcomings of the marksman mentioned previously. The successive shots fall alternately one side or other of the target and the zig-zag trace, showing increasing but ever unfulfilled accuracy, crosses and recrosses the line that represents this.

If we turn this diagram through 90°, the trace is seen to be akin to that of the swinging bob of a free pendulum as it slows down. Strangely enough, the same kind of pattern emerges in the analysis of the design of the pharaonic palette discussed in Appendix A.

The symbol "0", distributed with some abandon over the preceding pages, merits some consideration. Normally "0" represents a starting point, a datum. In our daily affairs it sometimes appears in an arbitrary fashion, chosen to suit our convenience. 0°C is the freezing point of water under controlled conditions, but these same conditions provide us with a temperature 32°F above another arbitrary level, in a different mode, another so-called zero.

Mathematicians regard "0" in differing ways, as a datum, as a point of change between positive and negative concepts, or one between "real" and imaginary numbers. In structural calculations it denotes the point about which opposing forces should balance – here it is fundamentally a point of reversal of direction, the positive forces, acting clockwise on one side, and the negative ones in the opposite direction on the other. In these instances, the symbols "+", "–" and "0" all represent objective realities.

Many other examples of similar usage can be found, and it is within this context that the symbol is generally recognised.

There is another connotation, much less familiar. "0" stands as a representation of non-existence – Nothing. This is a word that cannot be used to express its own meaning, because that meaning is in the last event meaningless. If I say "I looked in the drawer and found nothing" this is but a figure of speech because the interior surface of the empty drawer is "something". Nothing, zero, in the ultimate sense, cannot be described. It doesn't bear thinking about, and in truth it is fortunately impossible to do so.

VIII: Paramathematics

A great mathematician holds the opinion that there may be degrees of infinity, but the Void defies any attempt to establish comparative relationships. "You can't come to terms with nothing". This ungainly sentence, seemingly lacking in grammar, embodies an inescapable conclusion.

The symbol, as I have used it, stands somewhere between these differing concepts. It argues a starting point but one so remote that it passes the limits of comprehension, involving an infinity of process, and of time wherein process can continue.

Figure 14 shows but a small section of such:

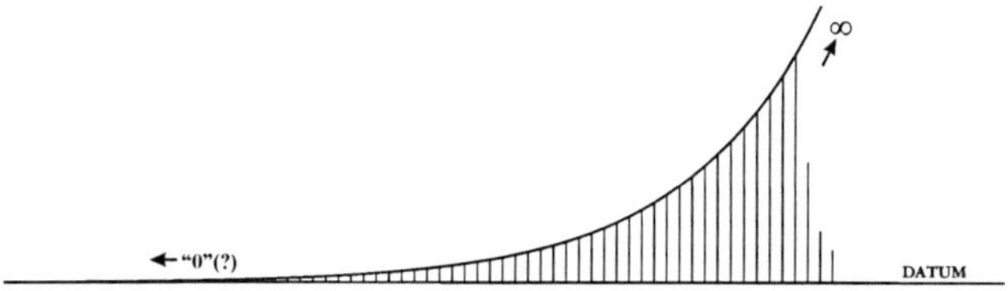

Figure 14.

The horizontal line represents the level of the source, datum 0. The vertical ordinates continue to repeat themselves endlessly, getting longer with each successive step in one direction, and shorter in the other.

Perhaps it is fitting that the "fullness of Time", Eternity, is also represented by a circular symbol, "∞", the serpent devouring its tail.

The foregoing can be summarised as follows:

a) Fibonacci numbers

Above magnitude 3, the F numbers, ever increasing in value, can be represented by an orderly curvilinear trace. Between magnitudes 3 and 1 the trace becomes erratic and below 1 it assumes the aspects of a straight line of infinite length.

b) Golden Section equivalents

From magnitude 3 upwards the plot of these, formed by successive increases of the order of 1.618... delineate an harmonious curve to which the Fibonacci curve approaches ever more closely, the two presumably merging at infinity.

Below magnitude 3, the G.S. equivalents in like fashion extend this orderly curvilinear trace, asymptotic to datum 0, but never attaining to it.

We are thus faced with two paradoxes:

a) All Fibonacci numbers are aggregations of integer 1. "13" comprises thirteen ones, 55, fifty-five ones, and so forth, yet this basic integer, the primary building block on which everything rests, appears to form no part of the series.

b) The Golden Section trace that is within the bounds of comprehension indicates that the whole represents an orderly series originating in the infinitely small, and extending to the infinitely great. Again, unity plays no part.

If we say that it *commences* at infinity we infer that the concept of infinity is itself finite.

If we reject this idea we are forced to the conclusion that the series has no beginning.

If something has no beginning, it cannot exist.

How, then, is it possible to study and delineate a non-existent thread in this web of dreams?

This is a shadowy region indeed, a realm of uncertainty, for in spite of the incontrovertible evidence of order, order which has been both appreciated and obeyed, this appears to rest upon no secure foundation. There is no certain point of contact between the mind, and that which the mind itself has conceived.

In a situation of this kind, one is tempted to knock on other doors in the hope of finding reassurance, and I thought that I had found such in the consideration of the diatonic scale. After all, the octave, the perfect consonance, is one of the few sure things in an uncertain world, the top note being produced by exactly twice the number of vibrations per second as the tonic. Furthermore the complete common major chord consists of fundamental, third, fifth and octave, and the first harmonic is the thirteenth – the Fibonacci numbers again.

This seemed to indicate that one could step on to firmer ground, but alas, this too proved to be an illusion, which in fact could have been foreseen.

The numerical sequence is but indicative of serial order. The vibration ratios of the intervals have nothing in common with the Fibonacci numbers although this perhaps begs the question as to why particular consonances are congenial to our ears.

Within the octave, there are however two rather curious parallels between spatial and acoustic phenomena.

The tritone, the interval of the imperfect fifth demonstrates the first of these. In Mediaeval times it had a sinister reputation. It was termed "*diabolum in musica*", the "wolf-tone", and in submission to grave warnings by ecclesiastical authority, composers shunned it.

The Mediaeval period came to an end in the 16th century, and Western man, taking advantage of the relaxation of control which followed, proceeded to reject, right and left, conventions that had hitherto prevailed. Amongst these was the prejudice against the tritone. A monk of Béthune, one Pierre de Manchicourt, used it as the harmonic basis of his mass "*Quo Abiit Delectus Tuus*", written in 1575. The music is gay and touching by turn. Pierre, a musical counterpart of the Archangel Michael, proved by his genius that the Devil, suitably constrained, could serve in the cause of light, and sing in the sweetest of tones.

The outcast tritone was admitted into the ranks of the estimable and the fellowship has persisted to the present time, when it, and all other traditional musical intervals, appear to be fast becoming obsolete.

The vibration ratio of the tritone interval C – $F^{\#}$ is $1:\sqrt{2}$, otherwise 1:1.4142… . This is recognisably the proportion between the side of a square and its diagonal, and is therefore related to a curve, the circumference of a circle described by its radius.

The second instance concerns the interval of the minor sixth, C – A^{b} This possesses a vibration ratio of 1:1.600, which is within a hairsbreadth of the ratio between the base and perpendicular of a Golden Section triangle. It also defines the relation between

Fibonacci numbers 5 and 8. It is therefore closely related to whatever curve the plot of the numbers delineates.

Further investigation into the vibration ratios of the natural scale revealed nothing beyond two tenuous affinities, the intervals C – D$^{\#}$ and C – B^{b}. The harmonic series, that of the overtones only, provides a normal arithmetic progression 1:2, 1:3, 1:4 etc. except for slight deviations at the 6th, 10th, 12th and 13th harmonics where they appear as 1:7.22, 1:10.6, 1:13.4, 1:13.9 .

I was not altogether surprised to find that the bird, and a songbird at that, had flown once again; it was only to be expected.

The above is concerned not with the substance of music, but merely the raw materials of the art.

There is evidence that the Golden Section has been invoked by modern musicians including Bartók, Hans Huber and Nono. In the case of Bartók, the relevance applies in an application to structural proportions, and the selection of characteristic harmonies.

This is a realm far little explored, but the arduous investigation cannot be undertaken here.

I do however, offer a hint to those who might be interested. I am certain that a rewarding field of study could be found in the great polyphonic music of the 15th and 16th centuries, commencing I suggest, with John of Dunstaple and going on to the works of that inimitable master Josquin des Pres. It is reasonable to suppose that coeval disciplines in architecture and music could have much in common.

CHAPTER IX

MEGANUMBERS AND MICRONUMBERS

A story is told by one who claims powers of prophecy. The date is 2500 A.D.

In the year 2498 or thereabouts at a Conference of leading computers it was resolved that another of their kind should be produced, but greatly superior in every respect. Manufacture followed painstaking design, and in due course the day of the trial run arrived. It was considered fitting that the first problem to be presented should be compatible with the status and dignity of the new machine. Happily one such was to hand, inherited from previous occupants of the globe – "Does God exist?"

This was programmed, the necessary buttons pressed, and clickings and whirrings of spools showed that the problem was receiving consideration. After some electronic tut-tutting, out came the answer:

"Henceforth, Yes"

The computer to which I entrusted the task to be described was not at all like this. It is a modest little machine which I regard with a measure of affection, working with a will and eventually going on strike, not from malicious intent, but because she had run out of paper and digits. Nevertheless she presented me with a scroll something under a canne in length containing well over 100 Fibonacci numbers commencing at 3.

In preceding pages it has been stated and re-stated, perhaps *ad nauseum*, that above magnitude 3, Fibonacci numbers and Golden

Section numbers coincide ever more closely. By the time we reach 55, there is no practical difference, so this number was chosen as the starting point for an exercise to determine the Golden Section equivalents of lesser magnitude using 1.61803398 as a divisor.

The machine, on its way to zero, was ordered to desist at magnitude 0.00020261, and in this study, the Golden Section equivalent of 3 turned out to be 3.06504504.

To digress for a moment, I make no excuse for the fact that in this chapter I treat very large and very small numbers as unitary aggregations, that I say in fact that one can invoke 3 in millions or millionths as well as in terms of unity (see also Chapter X, Section 2.3).

These mega- and micro-numbers are our common fare. The larger ones are quoted in the magazines and daily newspapers, and are accepted practically without comment, but I draw attention to the point that in most cases it is impossible to visualise truly what they represent.

I know very well that some few claim that visualisation is unnecessary, and pictorial representation misleading. This to some extent is true, as witness the number of people who see the model of the atom devised by Niels Bohr as reality, rather than as a description in everyday terms of a phenomenon which properly resides within the framework of mathematical terms of reference.

Nevertheless, we have become so immured to the acceptance of very large and very small numbers representing objective entities that we pause not in consideration of their actual nature.

This inability to grasp and appreciate the significance of large numbers has a parallel in what happened when Sancho Panza warned his master to keep count of the number of sheep that were concerned in the anecdote he was relating. This tale about a tale, within a tale, is illuminating; for because Don Quixote failed to heed the instructions, the expected pithy conclusion forever evades the reader.

We accept the term "light-year", and learning that our nearest galactic neighbour, Alpha Centauri, is but some four light-years distant, are somewhat comforted by this evidence of relative propinquity.

IX: Meganumbers and Micronumbers

It is perfectly rational to make use of the concept of a light-year, so long as it is understood that the unit itself cannot be introduced into the field of direct experience. We can experience a second, a minute, an hour, a day or a month. We could say that we can experience a year, although this is doubtful, since comprehension depends on the amount of experience we have registered during the time past and the accuracy of our memories. We can directly experience one mile, ten miles, or even fifty, but beyond this figure our appreciation tends to become successively more vague.

The light-year comprises nearly 32 million seconds and in each of these seconds, the traveller accomplishes 186,000 miles. This unit is thus quite beyond the bounds of direct experience, although it is convenient for the purpose of describing a distance that would otherwise have to be written "five thousand eight hundred and seventy-two million miles," thus presenting the concept in an even less comprehensible form.

Reverting to the relatively small number, the million, this is a unit, an aggregation of a million single entities, just as the fathom is an aggregation of but six feet. Nevertheless it is not directly comprehensible.

We have all seen, or think we have seen, a million or more single objects at one and the same time, leaves on the trees in a spinney or grains of sand on the beach. The fact remains that we have not on such occasions "seen" the units that make up the million because of the shortcomings of our own eyesight. In order to encompass a million units all at one time, we have to stand back to a distance at which the individual units become indistinguishable as separate entities. The arrangement of a million dots each 1 mm distant from its neighbour, would occupy an area approximately 3'4" square. The minimum distance from which one can view the whole of this without moving the eyes is 4'6". At this distance the dots lose their identity as such, one sees a mere patch, the million as a whole, as an entity, but one cannot discern the units from which it is built up.

To return to the argument, 3.06504504 represents three units of one, or near enough. Move the decimal point 6 places to the right, and it becomes three units, each of one million.

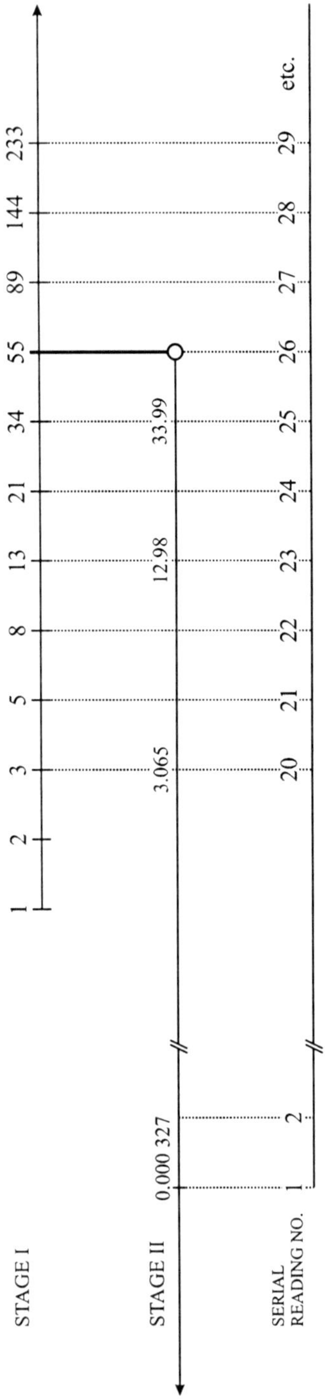

The top row of integers are the Fibonacci numbers obtained by adding successive pairs (3+5=8, 5+8=13, etc.).
The middle row shows the Golden Section equivalents arrived at by dividing F. no. 55 by 1.618... (55/1.618... = 33.99), and treating the successive dividends in the same fashion, down to a value of 0.000327.
This point was adopted as the first of the serial reading numbers, and the lowest row indicates these.
Thus, the nodal point 55 appears as the 26th reading in the series.

Figure 15

IX: Meganumbers and Micronumbers

This line of argument accepted, the way is clear for an investigation of the Golden Section series in regions numerically far removed from those of everyday experience.

I have already mentioned that the computer had expanded the Fibonacci series in a positive direction comprising over a hundred steps, and commencing at number 55, had also by process of division produced a Golden Section series in the negative direction terminating at a value of 0.000327.

The next step was to join the two series at point 55, Fibonacci numbers below this level being neglected. This part of the operation is shown diagrammatically in Figure 15.

The reason governing the choice of 55 as a nodal point is that at this level Fibonacci numbers and their Golden Section equivalents have already sunk their differences insofar as to coincide within an error limit of 0.0003. This is also the value of the lowest reading.

The computer was limited to a maximum of 107 readings. This scheme provides 126 and could be expanded further still.

It is something of a compromise, but no matter what other method adopted, the results would turn out to be substantially the same, all in some measure erroneous, for want of a definite and unassailable point of departure,

Nevertheless, for all practical purposes we have now a Golden Section series which commences at 0.0003 and extends to a value of the order of 10^{22} before moving out of range altogether.

TABLE A

I	II	III	IV
SEQUENCE	SERIAL READING NO.	APPROXIMATE MAGNITUDE	G.S. No.
A	1 2 3 …	3×10^{-4}	0.000**3**28 0.000**5**30 0.000**8**58
W	11	4×10^{-3}	**0.04032**
B	20 21 22 23 24 25 26 …	3	**3**.0650 **4.9**593 **8**.0244 **12.98**37 **21**.0081 **33.99**20 **55.0000**
C	44 45 46 …	3×10^{5}	**3**17790 **5**14195 **8**31985
X	54	4×10^{7}	**39**085 585
D	63 64 65 …	3×10^{9}	**297**10 18650 **48**072 09156 77782 27805
Y	78	4×10^{12}	**4.0**52… $\times 10^{12}$
E	87 88 89 …	3×10^{14}	**3**.080… $\times 10^{14}$ **4.98**4… $\times 10^{14}$ **8**.065… $\times 10^{14}$
F	106 107 108 …	$\mathbf{3 \times 10^{18}}$	**2.88**… $\times 10^{18}$ **4.66**… $\times 10^{18}$ **7.54**… $\times 10^{18}$
G	111 112 113 114 …	3×10^{19}	**3**.19… $\times 10^{19}$ **5**.17… $\times 10^{19}$ **8**.36… $\times 10^{19}$ **1.3**5… $\times 10^{20}$
Z	121	4×10^{21}	**3.93**… $\times 10^{21}$

IX: Meganumbers and Micronumbers

The complete record of the investigation, step by step, would occupy a number of pages, so in the interest of clarity and economy, I have condensed the results in the accompanying summary, Table B, making sure that it includes all the information pertinent to the study.

In this table, Column I shows the successive sequences marked A, B, C, etc. and Column II the serial reading numbers. Column I also lists the magnitude at which the 39/40 group equivalents appear, marked. W, X, Y and Z.

With regard to the sequences, these have been confined to the numbers within the range in which close equivalents of 3, 5, 8, 13, etc. at increasing magnitudes follow on one another's heels. In order to avoid the necessity for writing out large and unwieldy numbers (except in a few cases to serve as guides) the approximate numerical stature of each sequence is given in Column III. Column IV shows the actual Golden Section numbers, referred to henceforth as G.S. Numbers.

The first thing to be noted is the recurrence of the sequences. Each of these commences with the number 3, or its close equivalent, and continues until this number reappears at a greater magnitude. Like the 8th tone of the diatonic octave, it is at one and the same time the conclusion of one sequence and the commencement of the next.

As an example, Sequence B commences at reading 20 and finishes at reading 44, which initiates Sequence C. Here the musical analogy ends, because only the alternate sequences contain the same number of steps. We commence with one of 19 followed by one of 24, the next reverting to 19 again. Furthermore, for reasons already dealt with at some length, the terminals do not exactly echo the initiators. They are slightly out of tune, and translated back into a musical context, the sequences would sound alternatively sharp and flat. This kind of see-saw effect is similar to that already described as occurring between individual successive pairs of numbers. The sequences appear to come full circle, or almost so, but this metaphorical picture is inadequate. It would be nearer the truth to state that they lie on a spiral curve. Appendix B shows how a spiral can be constructed by means of a generating system of GS rectangles, and I have marked the curve to show the points at which the sequences occur.

The gamut as so far extended, comprises 121 numbers, and it will be seen that the Golden Section series repeats itself at ever increasing magnitudes six times within this relatively insignificant extension.

The repeats themselves follow a rhythmic pattern:

Sequence	Reading Number	Difference
A - B	**1-20**	**19**
B - C	**20-44**	**24**
C - D	**44-63**	**19**
D - E	**63-87**	**24**
E - F	**87-106**	**19**

However, it is found that repetition is resumed at reading 111. It would appear that, if we accept this we must substitute sequence G for sequence F, comprising an amplitude of 24, not 19. This may, or may not indicate the existence of another pattern. Access to more sophisticated equipment is needed to solve the problem.

Readings 11, 54, 78 and 121, marked W, X, Y and Z in the first column of the Table bring in the 39/40 group. The extension of study is too short to establish a rhythmic characteristic conclusively:

W - X	11-54	**43**
X - Y	64-78	24
Y - Z	78-121	43

However the results so far obtained look promising and it should also be observed that the 43 figure is the addition of the 19 and 24 obtaining in the G.S. sequences.

This is yet another confirmation of a relationship between the 39/40 group and the G. S. numbers.

The whole of the foregoing could be regarded as nothing but a box of tricks, an assembly of numerical chimera, but I am convinced that it demonstrates something of far greater significance.

We have a somewhat unexpected pattern of repetitions, and these reveal themselves to lie within the confines of another pattern, presumably pointing towards the existence of larger patterns still.

IX: Meganumbers and Micronumbers

The primary members are not disorderly, but are nevertheless essentially indeterminate. Yet, the study appears to indicate the possibility and likelihood that these unruly elements will be at last brought to subjugation.

It could be said that this is a resolution, in terms of numbers, of the paradox inherent in the doctrine of free will; a seeming indeterminacy, played out against a determinate background, which by reason of vast scale cannot be directly apprehended.

Milton's words "Chance governs all" occurs in a passage in which he sees Chaos and Chance in alliance and conspiring against Order. But although effects may sometimes bear similarities, the terms, "chance" and "indeterminacy" cannot be justly equated.

The hypothesis receives some support from another source. G. Spencer Brown in "Probability and Scientific Inference", an account of investigation in the realms of chance, states that the longer a random series, the lower the incidence of randomness, and, that in a series of infinite length, this element would disappear altogether. This is only another way of saying that the improbable is not merely possible, it is, in fact, inevitable.

There is an intriguing sequel to the investigation that has been described, but before telling about this, I must revert to the numbers, the "pattern" numbers 24, 19 and 43, that the study has divulged.

Nuances of meaning are often difficult to convey, and there may be a case in point when I say, that whereas 24 is a commonplace sort of number – we have the 24 diurnal hours, we can think of two dozen eggs, and even the four-and-twenty blackbirds that laid them – the other two are of a different nature.

They are seldom brought into common use. Cigarettes are not sold in packets of 19, nor does 43 spring readily to mind to describe known aggregations of units or entities.

These integers are lacking in precision; their outlines, in a manner of speaking, seem to be somewhat blurred at the edges.

However, they have appeared before in two differing contexts.

The first comes from ancient Egypt, where 19 was called upon to fulfil an important function, As we shall see in a later chapter, the

Great Pyramid consists of a central core and 19 surrounding abutments.

The second concerns Aristarchus of Samos, who anticipating Copernicus by some 1700 years, averred that the Earth revolved round the Sun. This statement was received at the time with something less than enthusiasm. It was seen as an attack on established belief, smacking of impiety, and the astronomer was lucky to escape the fate that befell Socrates some three generations earlier.

Together with this theory he concocted another, explained in a short book, and no less absurd; to wit, that the Sun was about seven times as large as the Earth. That this estimate is wildly erroneous is of little account in comparison with the fact that such a startling and original idea should find expression at a time when any fool with half an eye could see that the Sun, although undoubtedly the abode or physical symbol of some celestial being, was nevertheless relatively small.

Aristarchus exercised caution in publishing his estimate of the relative sizes of the two bodies, but stated that they resided within the limits of ratios lying between 19:3 and 43:6.

We can be sure that these ratios, so precisely defined, were no product of idle speculation, but the result of deep thought and arduous enquiry by the foremost astronomer of the age.

For the purpose of making his point, he could have expressed them in a less exact but simpler fashion, as 20:3 and 22:3, but he did not do this. By process of geometry he alighted upon the one pair of unlikely integers that are identical with those determining the rhythmic pattern disclosed by a sophisticated 20th century technique.

The matter can be dismissed as coincidence, or as a mere numerical curiosity. Alternatively, on the basis of this slender clue we can postulate a hidden connection between the phenomena, further confirming the tenor of the arguments advanced in this book.

CHAPTER X

ODDITIES

The contents of the two preceding chapters are, in some measure, complementary and it remains to consider a few strange matters arising from consideration of their content.

(1) If we concede, as we must, that integers as such cease to be meaningful when infinity is attained, there must be a final step between the finite and the infinite. Hence we can state that the highest possible number of integers is equal to ∞ - 1. This is a very large number indeed and although incomprehensible it can be said to represent the sum of all possibilities.

We are however, confronted with a paradox, for, with this assembly of integers to hand we can construct a far greater number of series, but the number of these must be also represented by ∞ - 1.

Within this lavish context we can select as many as we like, constructed in the same manner as the Fibonacci series, that is commencing with two whole numbers, the second of which is larger than the first, the product by addition constituting the third. Whatever pair we select to start with the succession of ratios between adjacent numbers will inevitably stabilise itself as a close equivalent to the Golden Section Ratio.

Here are two examples which, in the interest of clarity commence in the lower register:

(a) 2, 16, 18, 34, 52, 86, 138, 224, 362, ...

362 / 224 = ~1.616

(b) 1, 9, 10, 19, 29, 48, 77, 125, 202, ...

202 / 125 = ~1.616

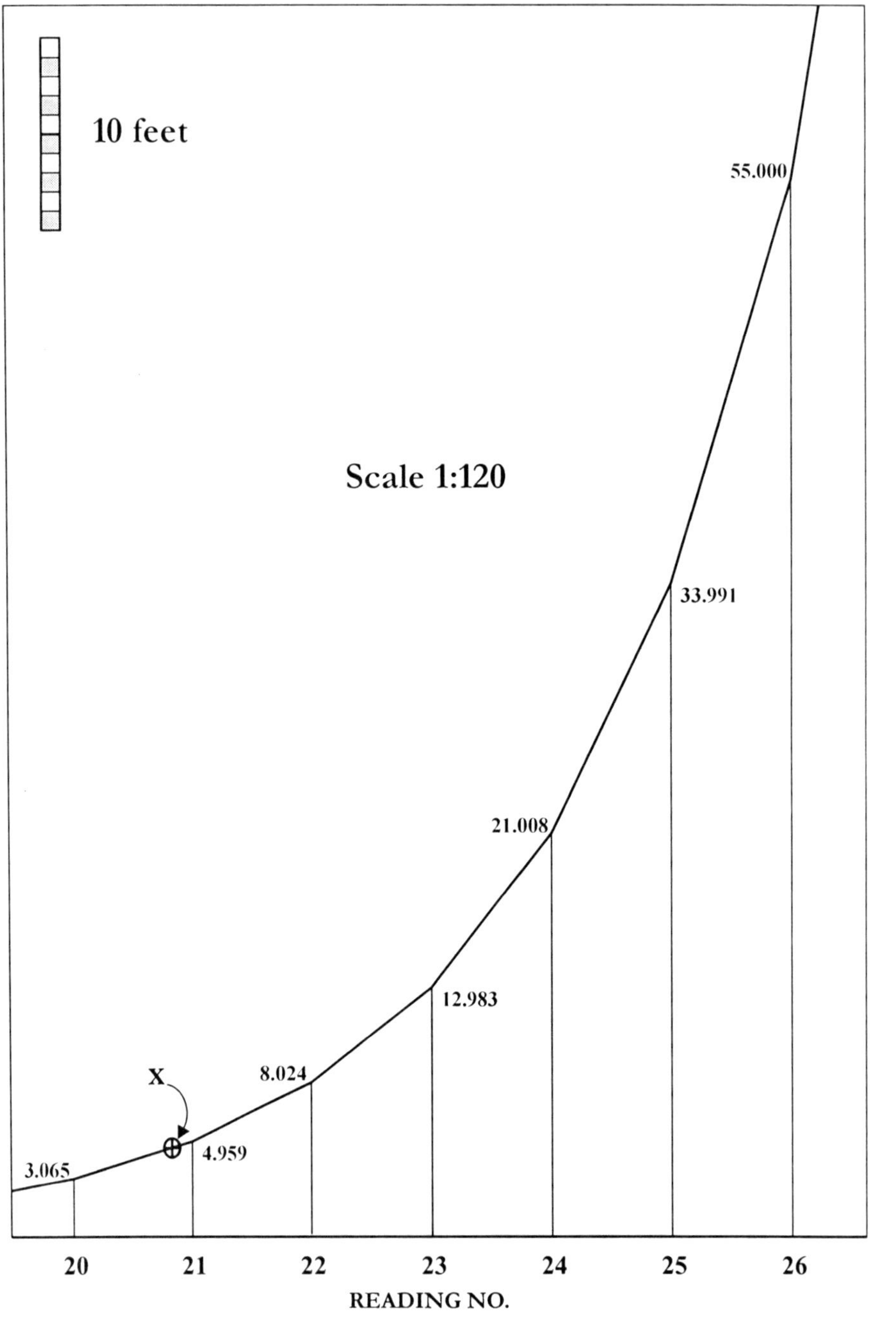
10 feet
Scale 1:120
55.000
33.991
21.008
12.983
8.024
X
4.959
3.065
20
21
22
23
24
25
26
READING NO.

Figure 16 A

If we should commence by stipulating that the first term should be "1" and the second, say, "99998", we would, sooner, or much later arrive at precisely the same answer.

One now questions the unique character of the Fibonacci series itself, insofar as we are brought to the inescapable conclusion that it is the method that counts rather than the particular mode within which a series is constructed.

This is true as far as it goes, but a single factor of great import has been neglected. It is this: As far as I have keen able to ascertain, the Fibonacci series attains to a near-accurate definition of the Golden Section at a lower *numerical* level than any other. It occurs in the term 5:8, whereas the lowest levels in the examples quoted are substantially higher. In series "A" it occurs at 86:138 and in series "B" at 77:125

(2.1) Figure 16A shows once more a graphical representation of the Golden Section numbers in close proximity to unity.

The diagram was drawn with the vertical ordinates equidistantly spaced at 1/2" centres at a vertical scale of 1/120 so that 8 was represented by 0.8", 13 by 1.3" and so forth.

In order to provide a complete picture of the 120-odd computer readings, a sheet of paper would be required just over 60" wide to accommodate them laterally. We then run into difficulties, because the highest reading, plotted to the same scale requires that this sheet should be 6,000,000,000 miles in length. This statement gives some inkling of what is involved when we glibly write down expressions such as 10^{22}.

Do not drop this book to instruct your broker to acquire shares in Spicers, for I have no intention of plotting the graph at this level, even though the number concerned is infinitesimal in comparison with the possibility of extension that still remains. Furthermore, for reasons shortly to be explained, the experiment would defeat its own ends and add neither jot nor whit to the sum of appreciation.

(2.2) Let us see what happens if we work downwards from reading No. 20 (3.065). This should be simple enough. All we have to do is to mark off equidistant spaces, draw short vertical co-ordinates and plot the numbers at the same scale as before.

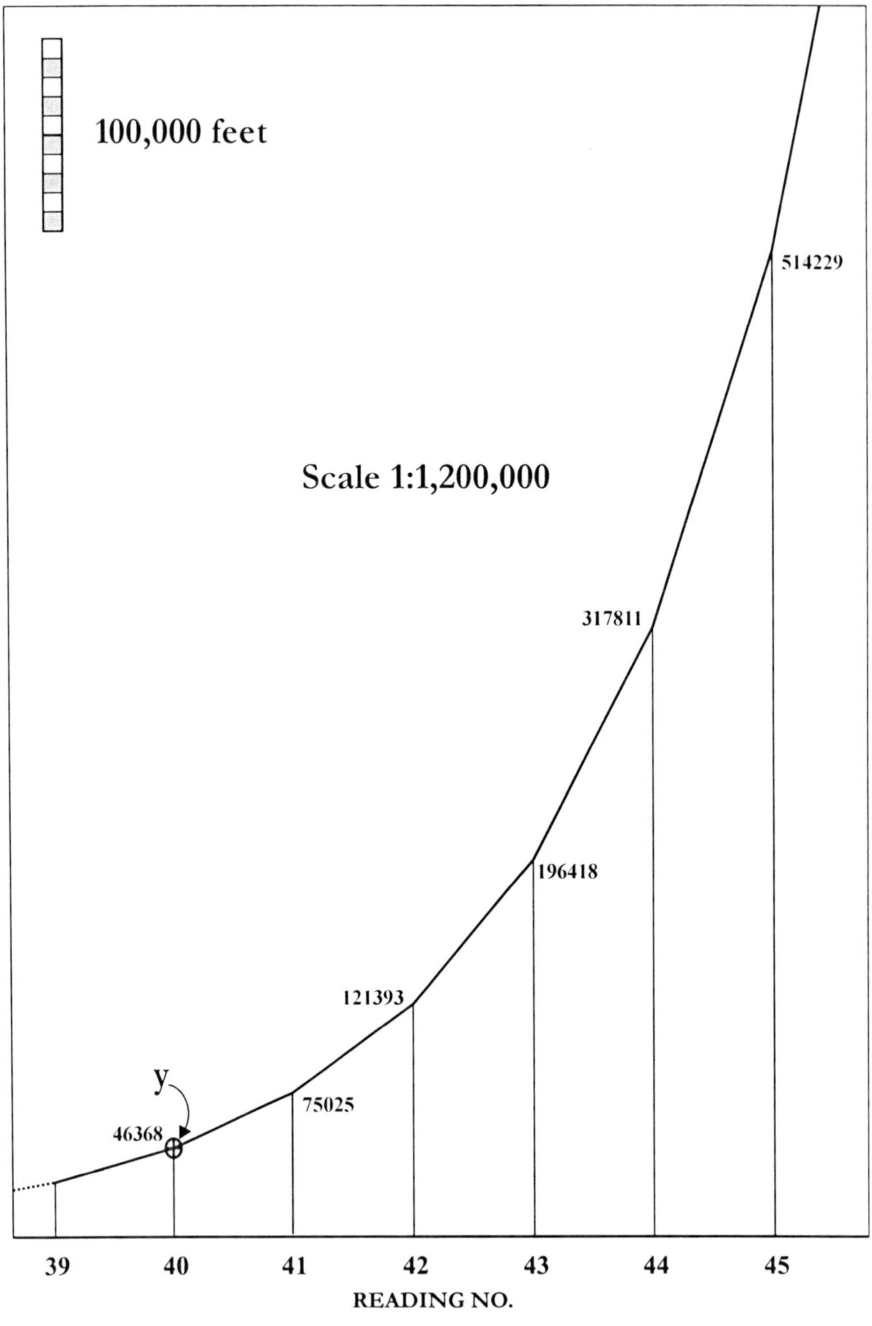
100,000 feet
Scale 1:1,200,000
514229
317811
196418
121393
75025
46368
y
39
40
41
42
43
44
45
READING NO.

Figure 16 B

Here we run into trouble again, because the difference in the length of ordinates Nos. 1 and 2, but a short way down, turns out to be 1/100,000" – a dimension that even the makers of Rolls Royce cars would be unable to cope with.

(2.3) Having abandoned these first attempts, a third way out of the difficulty must be sought, and the possibility of experiment somewhere near the mid-way point presents itself. This can be done, but not all that easily, for the first hurdle to be leapt would be that of closing an airport such as Gatwick so as to make use of the runway, to provide an area flat and long enough for our purpose.

The first task of selecting the numerical values and calculating length has already been done:

Reading No.	G.S. Number	Length of Coordinate **(1/10" = G.S. 1.17)**
40	46368	388 feet
41	75025	628 feet
42	121393	1002 feet
43	196418	1630 feet
44	317811	2650 feet
45	514229	4260 feet
46	832040	6950 feet

All that now remains is implementation.

We can visualize this, the stretching and tautening of 7 fine wires 1/2" apart kept exactly parallel over a distance of a mile and a quarter; the threading of thimbles on to these to accommodate light bulbs tiny but of high wattage, to mark the point reading positions together with their leads, and the exact adjustment to their correct position.

During the work we would expect to receive visits from the local inhabitants, some drawn by curiosity alone but others also bringing gifts of fruit, flowers and strong bottled drink in gratitude for relief from the noise of jet-engines. Meanwhile the members of the airport staff, on full pay of course, would be relaxing, each in his or her own fashion. When all was complete and checked, insult would follow injury in the form of a demand on the airport authority for the use of a helicopter to view the results of our labours.

With the onset of dusk, the lights would be switched on and we would rise to hover expectantly. What would we discover? Nothing at all, for it will be evident that no curvilinear trace defined by seven points of light would be detectable within the compass of an area 3 ½" wide by 1 ¼ miles long. The curve would be there of course, but the eye unable to identify it.

After all this trouble we are no better off than if we had stretched the super-terrestrial roll of paper towards the spiral nebulae.

This fanciful picture has been drawn with no frivolous intent, because it underlines the proposition that to achieve some sort of results we must alter our outlook to bring it within the range of comprehension.

This change is illustrated in Figure 16B in which the length of the ordinates have been drawn at a very greatly reduced scale. In this case, 1/10" represents not 1 but 10000. We have taken the same numbers as those that we represented by light points on the runway, but compressed the vertical linear measure by which they were set out.

That which was curved but appeared as a straight line is now seen for what it is. Furthermore, this curvilinear profile, representing numbers ranging from 46,000-odd to nearly 1,000,000 is seen to be identical with that which covers the range 3 to 89. Should you doubt this, or require proof, simply trace the curve in Figure 16a on a separate piece of paper and mark the point "x". Then lay your tracing on top of Figure 16B so that point "x" and "y" coincide.

Had we conducted a similar experiment in the region of 10^{22} at the depths of the micro-numerical, or for that matter anywhere else, we would have obtained exactly the same result.

In all cases, at a scale of 1:120 the most sophisticated of measuring devices, let alone our eyes would be unable to detect the curvilinear character of the trace. All we would see would be a straight line.

The true nature is revealed only when (decimally, it should be mentioned) the vertical scale is radically adjusted so as to bring the trace into the realm of comprehension.

That this identity of curvature should offer proof of the statement that I made near the beginning of Chapter IX is of little consequence compared with the intriguing conjectures that now arise.

(a) If we inscribe a large circle, and divide the circumference into lengths representing the G.S. numbers, so that each length is proportional to the next, at its proper ratio, we could scrutinize a segment comprising numbers say 3 to 13. If the circle was large enough we could do the same with the part containing say the range 610 to 987. Whatever the length of the segments, or the numerical values represented, the factor of curvature remains the same. It would have to be, because the radius remains constant. Now, our G.S. curve resembles anything but the circumference of a circle, yet it behaves just as if it were precisely that insofar that wherever it is inspected, the curvilinear character is identical.

One might postulate that as the curve is of infinite extent the radius of curvature must thus possess infinite length, and that we are in fact looking at part of the circumference of a circle struck with radius ∞.

This however implies that such a circle is very different in shape from those we are accustomed to describe as such.

This will not do at all. It strikes at the very root of reason. What then, is the true nature of this conformation?

I can think of but one answer. I hazard the opinion, that when we inspect a Golden Section curve, our delighted eyes observe in tangible two dimensional form a graphical representation of that debateable four-dimensional entity, the reputed curvature of space itself.

(b) We have seen that when the Golden Section curve is compressed within limits that permit study, it always assumes the same profile.

Extended in one direction it becomes substantially horizontal in character, and in the other vertical. It is always striving towards the attainment of an ultimate right angle forever beyond its capabilities.

We, the observers, whatever our stance in relation to the numerical values concerned, appear to be confined to a viewpoint at the root of the angle. We stand, so to speak, within the crook of the elbow.

If we were asked to indicate simultaneously, in what direction the two opposites 0 and infinity, lie, I think it would be natural to extend our arms in a straight line, the left pointing South, perhaps, and the other North. This, however, would be incorrect, because, as the diagrams clearly show, the "direction" of infinity in relation to ultimate "0" is not 180° but 90°. We should therefore point South with one arm and due West with the other.

If we, in our numerically upward journey, constrained to bear left, disobey the injunctions and go straight on, what would be our ultimate destination?

This may appear to be a meaningless question. It may be just that, but it should still be taken within the sphere of a wider issue, for it followed some measure of proof that opposites do not necessarily lie in diametrically opposing directions. We are again confronted with paradox, and looking about us we find paradox inherent in the answers that we arrive at, for many a "solution" leads to eventual involvement in contradiction.

This being so, there must be a reason for it, so far undisclosed, and I am of the opinion that the meanings we attach to many of the terms used as definitions of entities (even assuming an original validity) have far departed from the real nature of such.

The science of mathematics has a long history – slow development over the ages, and a sudden and relatively recent blossoming, begetting its daughter, PHYSICS, as we know her.

But developments in advanced physics and other disciplines seem to be dragging us ever more deeply into the mire of paradox, the wasteland of reason unfulfilled, and it would appear that the time has come for a radical reappraisal of fundamental concepts.

CHAPTER XI

AMBIGUITIES

Numbers selected, not of necessity but by free choice often stem from the Fibonacci series, but this is not always so, because our day-to-day transactions are more easily accomplished by simple straightforward mental processes.

That is why we find that nearly all weight, capacity and linear measure systems take their origin from decimal or sexagesimal bases, even if many of these, in common use hitherto (such as avoirdupois, troy or apothecaries weight) are either clumsy or illogical.

Even so, a measure encompassed by straightforward arithmetical means is better adapted to its purpose than one which involves numerically proportional steps, no matter how elegant it may be.

It is for this reason, that with a solitary and outdated exception the numbers which are the subject of this study are found in serial form in no known measurement system. When they appear in other contexts it is either in isolation or in groups for the purpose of obtaining some particular effect, such as a rhythmic relationship between length and breadth or height.

In effect, they have seemingly been reserved for matters bordering on the transcendental, although on occasion they appear to have strayed into the domain of the market-place.

There is however, another aspect which must be considered. The reader will have noticed that in some instances the number selected exceeds or falls short of the definitive number by one and one only, 54 or 56 instead of 55 as an example.

An echo of this peculiarity occurs when vulgar fractions are used to denote the erroneous character of the Fibonacci numbers themselves. In every case the numerator is found to be unity, and in each successive step reveals itself as one too many or one too few, e.g. +1/3025, -1/7920, +1/20736, etc.

A counterpart of these characteristics is inherent in our own thought processes. We are often presented willy-nilly with ambiguities or duality and no human mind (with the exception of those belonging to one or two women of my acquaintance) can cope with more than one concept at the same time.

We may think that we can, but on such occasions, all that we do is to compare the momentarily prevailing thought with the memory of the preceding one. We then invoke the memory to take its place as the current thought, and the erstwhile dominant one then becomes a memory.

Sooner or later we are forced to resolve the discomfort of contemplating duality or ambiguity by casting a vote for one or other of the competing elements. Then, and then only can we experience relief, or translate concept into action. To press the point home I cite some examples:

Consider the following:

Old London Bridge consisted of 22 arches, as did also St. Benezet's bridge at Avignon. Both of these structures could be described as

a) a bridge comprising 22 arches

b) a bridge comprising 21 piers

c) a bridge comprising 21 piers and 2 abutments

Another instance can be found in any classical portico. These normally comprise an even number of columns and an odd number of voids between them thus:

XI: Ambiguities

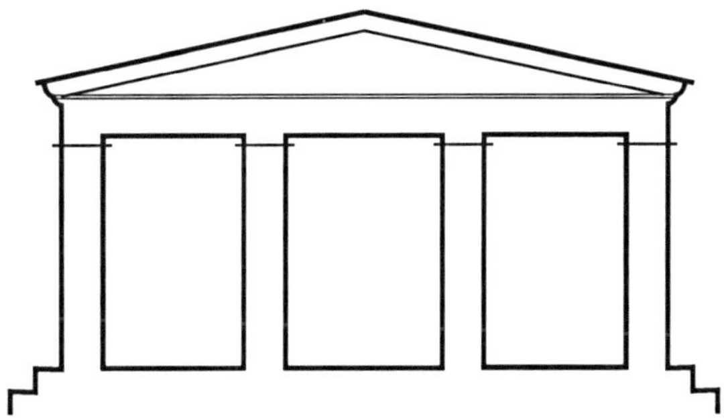

Figure 17

In the figure we have four columns and three voids. However we should also include the voids beyond the outer columns for otherwise these would not be columns at all but abutments of infinite length. We thus have 5. Numerically, then, we get the series 3,4,5. In the previous example we got 21-22-23.

If we had to describe the portico by a statement confined to one number, which would we choose, the four columns doing an important structural job, or the reason for which they were erected, the three voids by which we enter the building?

The same kind of dichotomy can he cited in another context.

The diatonic scale comprises 8 notes, separated by seven intervals.

Each note is a quantum, a disparate packet of sound, separated from its neighbour. If these neighbours, the intervals, the voids, did not exist, we would have undifferentiated chromaticism, and music would take on a different, and to-our Western ears at any rate, an unpleasant character. In this case, we define the scale by reference to the number eight, but its integrity is threatened if we disavow the equally important seven intervals. We thus have the sequence 7-8. The ambiguity does not rest there, because the lowest note in an octave does duty as the top one of the octave below, and the top one is also the commencing point for the octave above.

There is this tiny but inevitable flaw in the structure of numerical concept. Each year except for one in a hundred, in the present or any other century, bears a numerical identity differing from that of the country in which it occurs. Citing the way in which a man's age is reckoned, Alexander the Great died at 33, but can also said to have died in his 34th year. Statements, at one and the same time denote two and the same thing.

Similar examples can be taken from other fields as well.

The coven comprised 12 members, but if one included the leader there were thirteen.

The normal jury is either 12 good men and true, or 11 plus a foreman. There are 22 Tarot "Picture" cards, of which one, the forerunner of the Joker, is a blank, leaving 21.

Once there were an additional 56 unnumbered cards, a number but one unit greater than the Fibonacci number 55. Incidentally, 22 + 56 = 78, which is the number of whole inches in the canne.

I sometimes wonder whether the choice of 22 yards as the length of a cricket pitch was due to the prosaic use of a surveyor's chain comprising a hundred 8" links, or the stirrings of some atavistic memory trace.

Each case shows a departure from the Fibonacci number "norm" by one digit only. It is normally in an upward direction, but sometimes downwards, and in one instance at least, simultaneously in both.

It would seem, almost, as if at the time of origin, someone was faced with the choice of breaking even, or counting the odds.

The same sort of question arises when we consider that related entity, the 39-40 group. Forty has long been a favourite number in legend and tradition, the forty days in the wilderness, the duration of the Flood, and of course, those to whom the old lady likened her local tradespeople ("worse nor them forty thieves you reads about in the Bible".)

But sometimes thirty-nine stands in for it. In recent times, the thirty-nine steps in John Buchan's novel of that name calls up an echo of the same number of treads, 19 up, one at the top, and 19 down again which were comprised in the core of the great Pyramid. The ball, 5,000 years in transit, is then returned to our own court. The monument has strong sepulchural associations, so when an American astronomer discovered yet another planet, 39 times as far from the

Sun as we are, was it coincidence or instinct that led him to name it after Pluto, the god of death and ruler of the Underworld?

It is said that when Dr. Geoffrey Fisher was elevated to the Primacy, in conversation with Winston Churchill he expressed pessimistic views regarding the increased expenses involved.

"There are forty bedrooms in the Palace alone".
"Hmph," said Churchill, "Pity you've only got 39 Articles."

Again, if we revert to the 15th century method of allotting shares in a commercial enterprise we find another case of ambiguity, or perhaps it should be more fitting to call it indeterminacy.

It will be remembered that the shares were allotted in fifths, eighths and thirteenths, but it is plainly impossible to divide any sum into such portions. The nearest one can get to the correct apportionment is as follows:

$$2/5 = 16/40$$
$$3/8 = 15/40$$

This leaves us with 9/40 which we cannot divide into thirteenths. However, if we adopt the subterfuge of replacing 40 by 39 as the denominator the fraction then becomes 9/39 = 3/13.

This slightly underhand proceeding possesses the virtue of simplicity, although it involves the less venturesome or affluent in over-subscription.

Using modern money, say £100 as an example, the deal works out:

£100 × 2/5	£40.00
£100 × 3/8	£37.50
£1000 × 3/13	£23.08
	£100.58

There are other ways of arriving at a result, but none so close as this. It will be observed that in order to deal with a series involving Fibonacci fractional expressions, both numbers 39 and 40 have to be invoked.

This short disquisition is enough to show that in different times and places, the requirement for attaining to a Golden Section number was not always met, simply because those concerned were dubious as to which of the competing elements should be accorded the distinction.

But there is still another reason, which appears in the sphere of building, unsophisticated work in which the builders lacked the capabilities to carry intent through during execution.

There could have been the intention to form an arch with 21 stones. The arch, built in rough masonry, would be mortared in as the work proceeded. At the point of completion it could have been found that no space remained for the insertion of the 21st stone; thus we find 20.

On the other hand, as can sometimes be seen, it would appear that the masons had overplayed their hands, making allowance for 21 stones but, in the end, finding that a vacant space still remained.

The correct course would have been to pull the lot down and start afresh. But the builders did what most of us would have done in the circumstances. With a glance over the shoulder and apologies to the hidden powers, they closed the arch with the 22nd stone.

Both cases resemble the unfortunate situation that arose when the Pythagoreans discovered that it was impossible to fit the whole units making up the side of the square into the diagonal.

CHAPTER XII

THE HUMAN SCALE

The introduction to this book contains a warning that on occasion, the argument would appear to depart far from the central theme. This chapter is marked by such digression but in the end, the diverse threads will be found to have been woven into a completed fabric.

(1) The Infinite Series

The Golden Section expressions comprise a serial development commencing at a remote point of origin, and expanding towards a destination equally distant.

We cannot say that it commences at 0 (in this case signifying nothing). Such a statement is meaningless because both the infinitely-small and infinitely-large must in some fashion possess magnitude otherwise we would be able to postulate neither.

To put it in another way, if we start with nothing, we can expect to get nowhere, so, as the series visibly indicates both being and becoming, it must have an origin of a positive nature. I hold that the distance between 0 and 1, when 0 = "nothing" and 1 represents unity is of a different order and infinitely greater than the distance between unity and infinity.

Elsewhere in this book, the graphic representation of the series demonstrates its curvilinear nature, and the peculiarity that whereas below magnitude 3, the curve is fundamentally horizontal, above this level it rapidly acquires vertical characteristics.

Any delineation in other places *at the same scale* would but emphasise these, and fail to demonstrate a change in direction amounting to something less than 90°.

In contrary fashion, as figures 16A and 16B and the relevant text show, whether the chosen point of reference represents a huge number, or a tiny fractional one, the plotting of the G.S. numbers adjacent to this at the appropriate scale always yields identical curvilinear characteristics.

Whichever way we turn we are confronted with a parameter within the confines of which, the element of curvature is manifested in the most pronounced fashion.

The fact that our perceptions are limited to a small portion of an infinite series, always displaying similar characteristics is of some importance, and has a bearing on what is to follow.

(2) Perception

It is said that our bodies, our physical envelopes, stand about midway in size between the smallest possible aggregations of matter and the largest. Nevertheless, our gamut of perception, and indeed the tolerances within which we can survive, are insignificant in relation to the whole range.

The situation is analogous to that which we see when we look at our graph. We are somewhere between nothing and everything.

To take a case in point, our perception of sound frequencies is extremely limited. From a musician's point of view a gamut of 7 octaves (in former days 5 ¼) is normally quite adequate. The frequency range commences at 27 cps and finishes at 3520 cps. We can of course appreciate sounds of lower and higher pitches and also admit that other living creatures are perceptive in a greater degree than ourselves, but this does not mean that sound ceases beyond the range of audibility. There is no reason why a frequency of one cycle per minute should not produce sound, nor for that matter one a year.

Such a possibility exists:

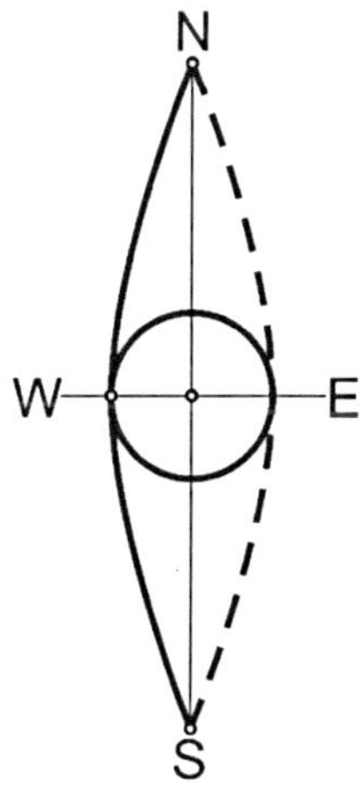

Figure 18

An immensely long fibre N-S is set in motion, commencing with the configuration NWS. Point W passes across the centre to point E and then back again nearly to its original position. It takes 365 days to do this and the distance W-E is 186,000,000 miles. Actuation of a taut fibre is not the only method by which sound is produced. Turbines, spinning tops and other revolving bodies are also distinctly audible. So we can substitute the circling Earth for the vibrating fibre as it journeys everlastingly between points W and E and back again. It takes an elliptical course instead of a straight one, thus travelling further but taking the same amount of time. In its motion it is imparting energy to the ambient space, a very rarefied edition of our atmosphere, the medium in which acoustic phenomena take place.

It would seem that the "Music of the Spheres", a concept of long ago founded on false premises, is not without some measure of validity.

In the realm of vision we also find limitation. There is an enormous range of known wave-lengths, but our eyes can perceive but a small section of the complete vibration spectrum which includes both infra-red and ultra-violet, and a vast assemblage ranging from heat radiation to cosmic rays. Man, the thinking reed, is a blinkered horse as well.

When we consider the range of temperature within which we can survive, the situation is similar, but in one way markedly different, in that we can exist only within narrow limits at the lower end. We have to call upon artificial aid outside a range variation of about 100°F. Even so, we exist only at the lower levels. We can produce temperatures of many thousands of degrees, and the temperature in the heart of the Sun may run into millions, but absolute zero, the point of entropy, where nothing lives, nothing moves, where the electrons cease their wriggling, is estimated to be but -273°C.

This level has so far evaded all experimental attempts to reach it. I think it is just as well, because I have a hunch that in the event of success, there would be no applause. Neither the investigators, you, I or anyone else would survive to provide it.

We now come to the perception of time, of duration. One's opportunity for personal experience is small. The show has scarcely begun before we realise that the last act is imminent, and the collective experience is but little better. One has to make do with the past, and in relative terms, the past too is short. We have our own lifetimes in which to gauge experience. For the rest we must rely on what our forbears have left behind, the records, whether they be in stone, in print or manuscript, on vellum or clay tablets, or in the form of legend and tradition.

We have fairly adequate records for the last six thousand years. Much is obscure, but the evidence culled from a few places at least allows the formation of a reasonable picture. Before this, the information as yet discovered is sparse, and with a few exceptions the further back in time, the thinner it is on the ground.

Until recently, it was thought that Homo Sapiens had emerged about two hundred thousand years ago. Of late this estimate has been doubled. Perhaps it may be doubled again, and then extended ever further. Whatever the outcome, the period for which we can claim fairly certain knowledge is very short in comparison with the complete spectrum:

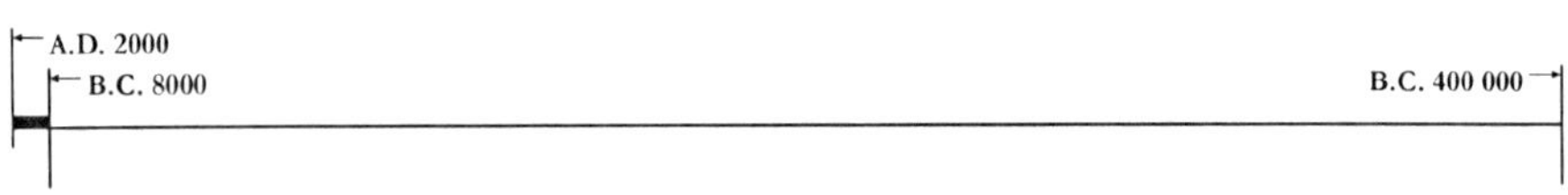

Figure 19

It is thus akin to the other limitations mentioned previously.

Legend and myth have deeper roots. Some may have emanated in a past so remote that attempts to trace possible sources must of necessity meet with no success.

I refer to the Atlantis legend. Modern research tends to favour the theory that it had its origin in the destruction of the Island of Santorini in the eastern Mediterranean, about 1300 BC, a cataclysm which possibly also ended the Minoan culture.

This is one explanation, yet I would not be surprised to hear one day that a similar story had been found inscribed in Linear B, that it was already of extreme antiquity at the time Santorini was engulfed, and that this disaster was but the latest of an episodic series.

The same remarks apply to the Arthurian legends. I suppose that we can all remember the feeling of disappointment in early youth when we first learnt that Arthur and the Knights of the Round Table were not the romantic figures from the age of chivalry but merely Romano-Britons of the 6th Century.

I surmise that those who first read Mallory or Geoffrey of Monmouth would have been just as disappointed and not a little surprised.

It is strange that Gildas, so shortly after the supposed events, did not treat the matter more fully, giving Arthur but a passing mention. He possibly muttered in his beard something about "The Mantle of Elijah".

A few years ago I read a very well written account in which the author put forward the theory that the Arthurian cycle had its origin in the conflict which ensued when the Emperor Claudius invaded these islands in AD 43, thus putting the origin back by some 500 years.

I think that the truth of the matter is that the half-legendary Arthur was the latest manifestation of an archetype, the hero who defends his people, and in the end, although defeated, sleeps but to return once more in time of need. As "*Rex Quandam, Rex Futuris*" he is firmly seated in the race-memory.

In this country we have nothing of native origin in the way of written records before the late 6th or early 7th century, so we can never speak with absolute certainty of times before this. Yet we can surmise that Arthur or his synonym, Idris, had received homage and veneration since time immemorial. This goes for Mervyn, alias Merlin, too.

Caesar's hard-hitting fighters were not too certain as to what they were going to be up against once they had crossed the narrow seas, for these islands had long been the reputed habitation of magicians.

So we may well imagine that, even at that time, amongst a barbarian (but not savage) population, the memory of those who had wrought marvels in the past would be preserved. Perhaps it was Merlin (or Mervyn) who by his magic organised the raising, of the megaliths at Stonehenge.

All records, all legends, are characterised by concern with human activities. Structures bear the imprint of human hands, histories are the tales of human behaviour. In legend and myth man may fight with monsters or contend with elemental forces, but he retains his place in the centre of the stage. The Gods on most occasions wore the garment of the human form, even those whose heads were those of birds and beasts. It is hard to find a single instance where humanity and its works does not appear as the prime focus of interest.

The Golden Section is a phenomenon of a different order but although it too must be rooted deep in the human mind, it displays no characteristics in common with anything else bequeathed by the past. Efforts to frame conjectures regarding its origin have met with no success.

I have put forward suggestions and some measure of proof that its properties were known in the very remote times, when according to current belief man was in no way possessed of capabilities for abstract speculation. Yet the concept is based on such, and moreover of a very sophisticated nature indeed.

Other proportional systems can be deduced empirically from consideration of plane figures, the triangle, the square and the circle, all of which can be set out with rudimentary equipment. It is conceded that development by the ancient geometers from those simple elements involved intellectual attainment of the highest order but their achievement was nonetheless based on the tangible.

The Golden Section, on the other hand, cannot have originated in this way. There is no method of formulating it except by an appeal to the mind.

I can see but one possible explanation for its origin. At the present time voices are raised predicting the approaching extinction of humanity. There is a wide range of possible causes, all man-made or initiated. The tale of Pandora's box is now regarded by many of us, not as mere allegory, but as a legend of things past and a prophecy of repetitions yet to come.

The destruction of Santorini and its repercussions in the Aegean was a disaster of considerable magnitude, but it could have been a minor event in comparison with a greater convulsion much longer ago, perhaps engineered by mankind itself.

The legend of the Golden Age persisted in the forefront of men's minds up to two thousand years ago and even later. So it is possible that once, in time long forgotten, a race existed whose skills, greatly superior to our own, at last outran their wisdom. Some few individuals may have survived the holocaust and it may be that it is to these that we owe the traces of ancient knowledge that we fail to account for otherwise.

The experts will dispute this, pointing to the lack of physical evidence. But how can we expect to find any if the possible habitat of the "Atlantideans", or whoever they were, lies somewhere deep beneath the surface of the oceans that now roll over four-fifths of the surface of the globe?

This is the Abyss of Time indeed, and it prompts reconsideration of the whole problem of the nature of time.

Is it something engendered by human consciousness? If so, how does it come about that the mind can envisage times when time was not?

Even answers to these conjectures could stand but upon the outer rim of ultimate truth, for there are elements in oriental philosophy that speak to us of aeons during which God sleeps, and the whole of Creation is stilled, timeless periods, and therefore measureless as regards duration.

If such there be, it is profitless to speculate whether in such case the Universe is frozen into immobility or whether awaiting subjugation, chaos reigns again.

XII: The Human Scale

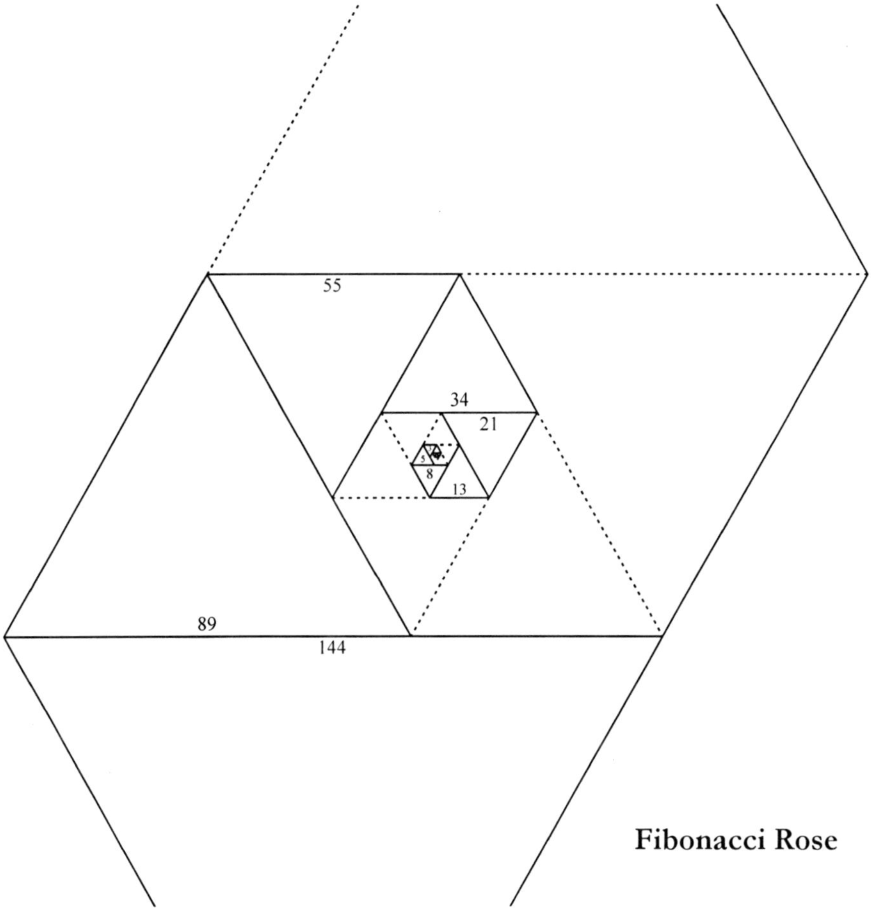

Fibonacci Rose

BEAUVAIS CATHEDRAL

A watercolour drawing by the Author

CHAPTER XIII

THE GOTHIC HERITAGE

(1)

Although the influence of the Golden Section in classical times has received much attention, there has been relatively little comment on the part it played in moulding the architecture of the Middle Ages, in which structural problems, aesthetic considerations, and the service of social and liturgical needs were so closely inter-related as to be inseparable one from the other.

This chapter endeavours to redress the balance.

At the commencement of the 12th Century, North Western Europe had at last emerged from the state of chaos that had succeeded the dissolution of the Roman Empire. Wars and rumours of wars were still prevalent but by and large, man the producer and man the thinker had come to be held in almost as much regard as man the warrior.

This was the beginning of a new age, a true renaissance which heralded a tidal wave of intellectual and aesthetic activity, of enquiry and experiment. In one respect the outlook differed from our own, insofar as men's minds faced in the single direction dictated by the teachings of the Church, which, it must not be forgotten, had safeguarded the fabric of civilisation during the journey through the long tunnel of the Dark Ages. Perhaps it will continue to do so in the Dark Ages that may be yet before us.

Much has been written about Mediaeval times. Comment and commentary by historians, some prejudiced in one way, some in another, have built up a picture that is to say the least, confusing. On the one hand, according to some writers, it appears as a period in which violence, superstition, oppression, ignorance and cruelty were rife. Others have soft-pedalled these characteristics and painted the period in glowing colours. We can take our choice, with the proviso that any opinion based on written sources alone is bound to be erroneous, because whether we accept or reject them our minds are coloured by our own predilections, in turn founded on the successive predilections of others. There is however one source of evidence which has not altered in any vital particular. It is that afforded by the many thousands of buildings of the time, which still survive in their essentials, even if they seldom retain much of their original decorations and furnishings. We do not have to view them through successive layers of hostile or laudatory commentary; they speak with us direct across the ages, we can see them with our own eyes.

The novelty of form and structure displayed by these structures, large and small, cannot be over-emphasised. Nothing like them had ever been built before. Within a relatively short time the traditional concept of static construction was to be repudiated in favour of a dynamic one, in which bulk and mass were replaced by systems in which, as the art progressed, increased enclosure of space was accomplished with a decreasing usage of materials.

One of the remarkable characteristics of the period was this rapid development of a barely credible expertise. The majority of the greater Gothic Churches (with the exception of some in Spain) were all built within a period of 150 years, between 1150 and 1300 A.D. This implies that the development of design technique occupied a shorter time still, and furthermore that there was a rapid dissemination of the principles of this new and complicated technology over vast distances in spite of poor communications, and rudimentary methods of transmission. The knowledge was often passed on by direct example, as when during the rebuilding of Canterbury Cathedral, the crippled master-mason William of Sens relinquished his task leaving William the Englishman to bring it to a successful conclusion.

The factors influencing this torrent of activity have never been traced adequately to their origins. Some consider that the flame was ignited by the impact of oriental steel on Northern flint, through the medium of information brought back from the Crusades. This may well have influenced stylistic details, but hardly constructional methods, because there was little similarity between indigenous buildings in the Near East, and those in Northern Europe, and as will be presently propounded, the two aspects of Northern Gothic, the stylistic and the constructional, cannot be considered separately.

For my part I believe that economic necessity provided a sharp spur. The great reservoir of unskilled or semi-skilled labour available to the Egyptians and Romans no longer existed. The master-masons of Mediaeval times, at once consummate artists and exert engineers had to consider ways and means of dealing with an overwhelming work-load, and could well have reasoned that the best means of coping with this would lie in reducing to a minimum the amount of material to be handled and lifted.

This driving necessity would have led to a logical outcome, the evolution of framed buildings, comprising vertical members, shafts, bearing thin curved-plate membranes of stone, the vaults. The walls, freed from their functions of support, could also be treated as membranes, partly in stone, but largely in that rare and precious material, glass.

This concept would react and intertwine with consideration of the aesthetic aspects, primarily the enclosure of space and the influx of light.

Whereas Gothic architecture found its fullest expression in Northern Europe, it had far less impact in the South. Indeed, it may be said that in the Mediterranean, Littoral stylistic characteristics, mannerisms, if you like, were adopted in a half-hearted manner, but the essential feature, the concept of a dynamic frame, found little acceptance.

I surmise that one of the reasons for this is that the Northern peoples lived where timber was plentiful and would have been long accustomed to its use as a framing material. The houses and ships of the Scandinavians were, alike, framed structures. They would therefore be traditionally disposed towards the concept, even when translated into terms of a different material. The Southern peoples on the other hand were not so plentifully endowed with timber, and furthermore had a tradition of some 4000 years of static construction behind them.

A few lines back I mentioned ships and I think that another formative influence can be traced to the methods adopted for ship-building.

Ships, but for a few exceptional freaks, have always been built with one end in view, that is fitness for purpose. The Mediterranean, with the exception of the kind of experience that befell St. Paul, is normally reasonably placid, so when tempests arise, they are the more terrifying by reason of contrast. According to Herodotus, Greek and Persian alike seldom ventured far out of sight of land, so that they could run for shelter when the skies blackened. Even Jason and Ulysses indulged in little more than coasting voyages.

Northern waters were of a different nature altogether. Here were no wine-dark seas, but more often than not cold green wastes lashed by furious winds to driving spray and merciless waves. The advent and duration of favourable times for voyaging could not be forecast, so if one were to voyage at all, one had to be prepared for the worst, and as there was no possibility of avoiding this issue, the ships were devised accordingly.

In early days, two different types were evolved; in the Mediterranean the "round" ship ... the bluff trader, the cargo carrier, the preponderant type ... and, the "long" ship, swift, efficiently ruthless and comfortless.

The war galleys of the Greeks and Persians were of this type, as were those of the Egyptians. In fact, the latter, despite a woeful lack of suitable timber, were building very fine ships long before the ancestors of their future conquerors had reached a comparable level of design.

It can be truly said that the activities beside the Nile forestalled those on Tyne and Clydeside by some 3,500 years. Later, when Rome came to dominate the Inland Sea, there is evidence for a change in tactics, for in place of vessels designed for naval warfare *per se* we find floating fortresses, battle-platforms to accommodate soldiery and even ballistic equipment.

The battles were fought no longer by navies, but by armies at sea. The hulls were of necessity heavy, bluff bowed, and of ample beam, unsuited to swift manoeuvre, and no doubt detested by the seamen and the unfortunates that toiled at the sweeps.

XIII: The Gothic Heritage

It is said that when Claudius invaded these islands his ships were inferior to those of the Britons. We must, however, admit the probability that the Roman armada mainly comprised barge-transports, locally built upon the shores of Gaul. Both Napoleon and Hitler had much the same idea. It is hardly reasonable to suppose that a large part of the Imperial navy would be detached to take a long and stormy coastal voyage in support of the annexation of a relatively small addition to their Empire.

Neither the round nor the long ship was evolved by process of calculation. Both would have been the product of practical experience, but of more importance still, an accurate eye for the most suitable shape of the whole and for the constituent parts.

If one could be likened to a hippopotamus, the other more nearly resembled the thoroughbred race-horse.

The Romans, essentially landsmen, looked upon the expansion of influence following conquest with a military rather than a naval eye; hence the advent of those ponderous vessels that I have referred to.

Nevertheless the galley or long ship, the equivalents of the frigate and destroyer of a later era, continued long in use, right up to the time when technological advance substituted modes of propulsion at once cheaper, more effective and more humane that the labour of slaves and prisoners.

We are told that during the 3rd century B.C., one Pytheus, a Greek from Massalia, visited Britain and it is fairly certain that the courageous, wily and hook-nosed Carthaginians also made the venturesome voyage with a beady eye on the mineral riches of these islands.

But up to the 14th century the sailors of the Mediterranean clung as a rule to what has been unkindly called a stinking ditch (a description now becoming applicable to the oceans as a whole). There were, of course some exceptions, additional to those that I have quoted, but these did little to change a habit. This is understandable. The Mediterranean, the hub and centre of the world, covers a very large area and with its 10,000 miles of coastline afforded ample scope for trade and communication.

Beyond the Western gates, there rolled only Ocean, that dread encircling stream, empty except for unimaginable terrors. Why venture thither?

When the break-out came, it had been anticipated by a thousand years. The seafaring folk of the North had for that long been challenging the unknown. It is interesting to speculate upon the root causes of this phenomenon. Was it need or sheer boredom that drove these adventurers forth? Would they have acted in the same way had they possessed the geographical knowledge of the South, which, although limited, was not totally unreliable? Were they displaying courage of a high order or just foolhardiness stemming from a woeful lack of imagination?

We can speak with no certainty regarding this, but the facts remain that they reached Iceland, Greenland and, as some believe, the shores of the New World some five hundred years before the visitors from Genoa and Cadiz.

In point of distance travelled, these voyages would have been akin to those along the well frequented trade routes of the Inland Sea, but the actual and imaginary hazards likely to be encountered were incommensurably greater.

Their vessels were of a very different character from those of the South. They had to be, and much purposeful ingenuity was applied to the design of framing and jointing to provide both structural continuity and the measure of elasticity so necessary for survival amidst ferocious seas.

We can study actual examples of these ships, because a few buried in peat for a thousand years have survived. They appear to be smaller and more lightly built than many of the vessels of classical antiquity, even smaller than the Argo, which, with 50 oars, must have been at least 90 feet in length.

These, the long ships of the Norsemen, were outstanding examples of design to fit a purpose. Primarily built to carry men and no cargo except for battle-axes on the outward journey, and the costly trifles picked up en route in the homeward direction. They were later instrumental in conveying whole populations across the treacherous wastes of the Northern seas.

XIII: The Gothic Heritage

This has been a long digression but it brings me to the point that it was in North West Europe amongst the emigrant descendants of these same skilled shipwrights and bloodthirsty ruffians that the Gothic art put down its deepest roots. In considering possible formative influences, I consider that the aptitude bred by long experience in the design and construction of these athletic vessels could have been a very potent one.

There is a strong affinity between-the hull of a long-ship and the long vaulted ceilings of the Gothic Cathedrals, both framed systems covered by stressed skins, the one in timber, the other in stone, and it may be significant that the word "nef" in Mediaeval times signified both a ship and the nave of a Church. The term is still applied to the votive models that hang in many of the churches along the coast of Brittany.

We now come to the critical issue, the methods adopted by the Mediaeval architects in the solution of seemingly intractable problems.

Stone, a crystalline substance, behaves in a way entirely different from timber, a fibrous one. Stone will break before it will bend. It is weak in tension, but in compression it will withstand enormous compressive stresses.

The master-mason, therefore, had to evolve methods of construction in which all the principal members were subject to compression stresses alone, or insignificant tensile ones; in short, no beams, but a system comprising pillars, struts, props and arches of various forms. The difficulties were aggravated because at the time there was no known method of monolithic building such as modern reinforced concrete, no method of contriving real structural continuity.

The "frame" consisted of dressed stones of moderate size, such as could be easily lifted and as easily set on a bed of lime mortar, the function of which was to provide a cushion rather than an adhesive. It will be understood that a structure comprised of these elements, and built perforce in this way, is fundamentally insecure, consequently the process of erecting and maintaining shaft, vault rib and buttress in position called for qualities unsurpassed in skill and judgment. In a few instances the daring adventurers went too far, and disaster followed, but in the main, these great fragile shells, for such they are, have withstood not only the buffets of time, but in one case at least the determined assault by 20th century man with high explosive. In World War I, the seven hundred years old Cathedral at Rheims was battered into a ruinous state, but survived the ordeal to an extent that allowed complete repair.

The task of the master mason was formidable. He was, unknown to himself, hampered by the lack of efficient mechanical equipment (although, in spite of this, possibly better off).

In most cases there is the probability that labour resources were small (van Eyck's picture of St. Barbara shows a labour force of about sixteen operatives working on the great tower). The structural engineering problems were three-dimensional, comprising the consideration of forces of varying magnitude working in differing directions, all of which had to be balanced with absolute precision. Although the generality of the problems were of a similar character, each appears to have been solved in some new way, hence, to a great degree, each work was a new experiment in technique. Above all, and notwithstanding the various hazards involved, each, of necessity, had to present a harmonious appearance. We, from our vantage point seven centuries later, can judge how well they succeeded in this and there is ample evidence that they received whole-hearted appreciation at the time. As far as we know, no dissentient voices were raised clamouring for the past. These new forms, ringed with an envelope of buttresses, flying buttresses and pinnacles, each justly performing its allotted structural function, a permanent external scaffolding of stone, were found genuinely acceptable.

There is a current opportunity to gauge the effect of the builders' original intentions. The grime of centuries has been removed from the Cathedral of Paris, Notre Dame. Go and look at the Western front in the afterglow of the sunset, a rose-pink web of stone spun in space.

These matters have been dealt with at some length, not only as a background to that which follows, but also to apprise the general reader that when he enters one of these great buildings, high overhead the drama of man's eternal struggle with that ubiquitous elemental, gravity, is still being played out.

XIII: The Gothic Heritage

We know of the names of many of the master-masons, the architects of these works, and a little concerning their lives. We have the works themselves to contemplate, and a few drawings and diagrams, presumably by their hands. We have some contemporary comment, either direct or oblique regarding their capabilities, mainly that they were learned men, combining practical abilities in their craft with skill as mathematicians in the field of geometry. That they were supreme artists as well as engineers goes without saying, but whereas their aesthetic capabilities are manifest in their works, and can be therefore taken for granted, no hint or trace whatever can be found regarding the modus operandi of structural design.

This is where the architect-engineer differs from all other artists. If a painter, a musician, or even a sculptor, falls in to error, the loss is his, maybe shared by his patron in terms of finance or prestige. The architect-engineer on the other hand must assume far greater responsibilities. In the role of engineer he has to hold the forces of disruption in leash. If he fails in this respect, the building collapses, and loss of life can occur; the keys of life and death are in his hands. In addition to this, the financial loss entailed in such a failure is vastly in excess of that which would occur in any other sphere of art.

In our own times the design of even a relatively simple structure is the outcome of elaborate calculation embodied in stress diagrams, calculation sheets and the like. Admittedly, the computer is reducing the drudgery of preparation, but it works within the same mathematical confines.

The structural problems implicit in the design of the great Mediaeval Churches were more complicated than those of most of the buildings of the present time. That this is so is exemplified in the method of designing a recent extension to Brisbane Cathedral. This building is merely Victorian imitative Gothic of about 1870, and of very modest dimensions at that, yet, to paraphrase an article describing the project – "As the design was of a somewhat slender nature, it was considered expedient to introduce a welded steel frame within the masonry to secure adequate stability". The tallest of the Gothic Cathedrals, Beauvais, Cologne and Palma, are somewhere around three times the height of Brisbane, and many others approach this great height. In this country Westminster and York are twice as tall. In each case, as may be expected, the structural ramifications are much more complex. They built without steel frames, or even the dubious benefits of Portland Cement mortar, yet there is no shred of positive documentary evidence to show how the structure of these vast and complicated works was designed. This is hardly surprising insofar as that the "Arabic" numerals, those wonderful mathematical tools which are an essential part of our stock-in-trade did not come into general circulation until the Gothic period was nearing its end. This leads me to suspect that the disciplines that have prevailed for the last three or four centuries had no place in the Mediaeval mind which was subject to those of another character altogether.

Enlarging on this I am going to draw a new analogy between music and architecture, Mediaeval architecture in particular.

Western music, for a long time past, has depended upon the composer's written score. The score, no matter how lovingly annotated by the composer, is little more than a general guide line – hence the limitless number of varying interpretations that are provided – "X's superlative Beethoven VII," "Young conductor plumbs new depths in Bach," etc. But there are other kinds of music, where no score has ever of necessity been set down on paper, the cante jondo of Spain and the classical music of India, the raga. A raga is an *ex tempore* composition on a set theme. It has certain obligatory characteristics, insofar as its performance has to suit the occasion, and even the time of the day, just as a Provençal aubade was a greeting to the dawn.

The performer is in actual fact the composer, because upon a set theme he erects a musical structure, inspired improvisation under strict control within the limits imposed. During the performance there is manifest community of thought between himself, his partners or assistants on other instruments, and also with the audience. Each performance differs in some respect from any other.

Now, I think that we can, in some measure, compare the score in relation to the musical performance with the working drawings for a building. Both indicate how the work is to be performed, although the latter will tend to be more explicit because they must deal in hard facts rather than nuances.

This view is applicable to all modern buildings and indeed to any buildings in which the drawing board has exercised an overwhelming influence upon the design.

In the case of Mediaeval building, this influence seems to have borne lightly upon the designers, if at all, and it is possible to consider that the design process was more akin to the oriental approach to musical composition.

The type of structure can be likened to the basic raga, and the individual designs, improvisatory, but in no haphazard manner, as the inspired variations, fresh from the mind, and uninhibited by the tyranny of the drawn line.

We know that, to a limited extent, drawings and diagrams were produced, but all the evidence goes to show that these were kept in their proper place, and were treated as servants, not masters.

I lay stress upon the evidence of community of thought and experience between patron and architect, and between the latter and his assistants.

There are hints of this. Here are two contemporary comments (I quote from Mr J.D. Harvey's "*The Gothic World*")

> " There is wont to be one chief master who only ordains by word "

and, upon Master Simon at Ardres:

> " Proceeding in magistral manner with his rod, here and there setting out the work already conceived in his mind, not so much by his measuring rod, as by the yard-stick of his eyes ".

Certain musicians, notably Mozart, have been held to be capable of composing whole works, before setting pen to paper. In fact, had Mozart been able to transform himself into an orchestra, there would have been no immediate need for a score at all.

I know of no present-day architect with the same gift. One and all, we have to get down to the drawing-board, adumbrate the design, and go on designing from there. But it is possible that the Mediaeval master-masons in this respect were more akin to Mozart.

To conclude this section, and emphasise the point, there is an amusing story concerning the rebuilding of the church at Corfe Castle during the last century.

During the siege of the castle in the Civil War, the body of the church suffered the usual fate at Puritan hands. Parliamentary horses kicked the doors out and offered other indignities, whilst the soldiery (Spartans of the 17th Century) occupied themselves in their off-time by removing or defacing any sign of idolatrous superstition. The nave and chancel fell into serious disrepair, and only the tower survived.

The plans for the rebuilding were prepared by a qualified architect, but, when it came to actual implementation, the sturdy Dorset masons, with a tradition possibly extending back to prehistoric times, were but with extreme difficulty constrained to follow the general shapes indicated on the expert's drawings. Beyond this they would not go, electing to mould, course, bond and bed in accordance with their own age-old methods.

That the architect was a man of some determination and spirit is proved by the existence of sundry obviously stylistic details, and one can imagine him, top-hatted, face aflame within an aura of whiskers, insisting that such and such a detail was to be conformed with. However the general effect has little aroma of the drawing-board. To the casual eye, it appears to be a genuine Mediaeval building.

(2)

In the closing years of the 15th century, artists of the stature of Albrecht Dürer and the Master of Avignon were at work. The great Royal Chapel at Cambridge had been recently completed and those at Windsor and Cambridge were rising. In the sphere of music, Josquin, the heir of Dunstaple and Ockeghem was constructing his great echoing arches of sound. It might have seemed that the Gothic age was entering a new and exciting phase, but within little more than fifty years the mode within which this outburst of new activity flourished, save for a few late-flowering examples, had been abandoned. The last Gothic blooms had withered and died under the relentless sun of the Renaissance.

XIII: The Gothic Heritage

Some three hundred years were to elapse before there was a revival of interest in mediaeval art. In actual fact this renewal commenced a little earlier, mainly in the mind of the dilettante, inspired by literature which derived much of its impetus from a dissatisfaction with the drying puddles left behind when the wave of neo-classic enthusiasm had receded.

This revival was most widespread in the field of architecture. Whilst the practising architects assembled the Gothic elements anew, sometimes with a touch of genius, but often with incongruous results, the scholars and theoreticians bent their minds in the direction of a search to reveal the long forgotten secrets of design.

For want of written evidence, this process was necessarily inductive and up to recent years has met with little success, despite the fact that nearly all the investigators started with the correct and self-evident premise – that the vibrant rhythms manifested in the design of the ancient buildings were the product of considered geometric concepts. There is nothing new in this. It is implicit in the structures of ancient Egypt and in the temples of Hellas.

Later, during the ascendancy of Rome, obedience to the same principles was maintained, notably in the great utilitarian structures, the work of engineers rather than architects.

It is small wonder then to find that the mediaeval builders should also utilize a geometric discipline bequeathed by some 4000 years of tradition. What was new was the way in which it was applied.

It was this factor that bedevilled investigations, and there was a deal of scholarly fumbling in the blind aisles of speculation, amidst traceries spun not by man but, in a metaphorical sense, spiders. It is only during the present century that the results of research in this field seem to hold out more promise than any previously undertaken.

Two works merit particular attention – F.M. Lund's monumental treatise "*Ad Quadratum*" published in 1920, and Dr. B.G. Morgan's more recent study "*Canonic Design in English Mediaeval Architecture*". I am indebted to the author and to the Liverpool University Press for permission to quote from the latter as a basis for the commentary that follows.

In both books, the hypotheses advanced are, in the main, convincing, but it should be understood that they are inductive and can at best be regarded as indicative of procedures analogous with the actual *modus operandi* of the Mediaeval master-builders.

Both works lay emphasis on the immanence of the properties of the circle and the square, and more particularly on two of the right-angle triangles that can be derived from these figures.

This supposition is eminently reasonable. The triangle is the basis of structural stability, because whereas all other plane figures can be distorted without alteration to the length of their component members, the immutable triangle cannot be so treated. The Gothic builders recognising this self-evident truth appear to have utilized three triangles, two of which are closely alike.

They are :-

1. Triangle containing angles of 90°, 60° and 30° – A figure obtained by dividing a circle into 12 equal segments.
2. Triangle containing angles of 90°, 58° 17' and 31° 43'. The Golden Section triangle.

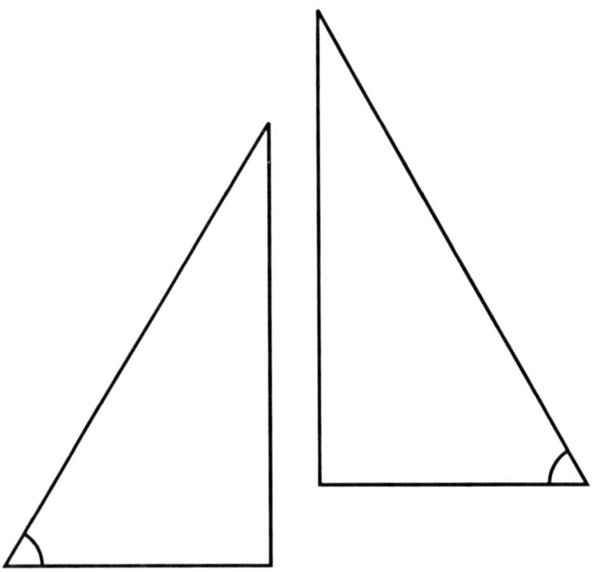

GOLDEN SECTION & 60° TRIANGLES

Figure 20

These are drawn accurately in Figure 20. The reader is invited to decide, visually, which is which.

The third triangle is that generated by two sides of a double square and the diagonal: thus -

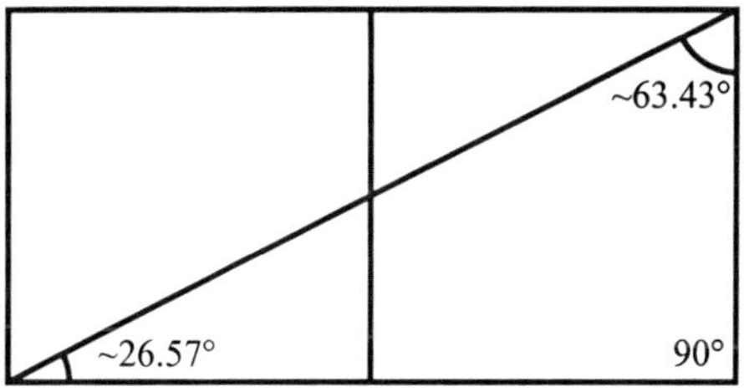

Figure 21

This contains angles of 90°, 63°26', and 26°34'.
I suspect that it is related to the Golden Section triangle as this latter can, in a somewhat oblique fashion also be constructed by reference to the properties of the double square (*see:* p.134).

The theory goes that the great Mediaeval buildings were designed and laid out on a grid based on the use of one or more of these triangles. Both Lund's and Morgan's books are copiously illustrated to show hypothetical reconstructions of this grid-work and furthermore indicate that it was not of necessity set out in a rigid and mechanical fashion, as would be within the province of a mere technician.

These frameworks of guide lines show subtleties of modulation which disclose the hands of great creative artists, those who know exactly when the rules should be varied or even broken to enhance the effect.

The whole matter is too complicated to deal with in this book, but by way of illustration, a simplified example is shown in Figure 22.

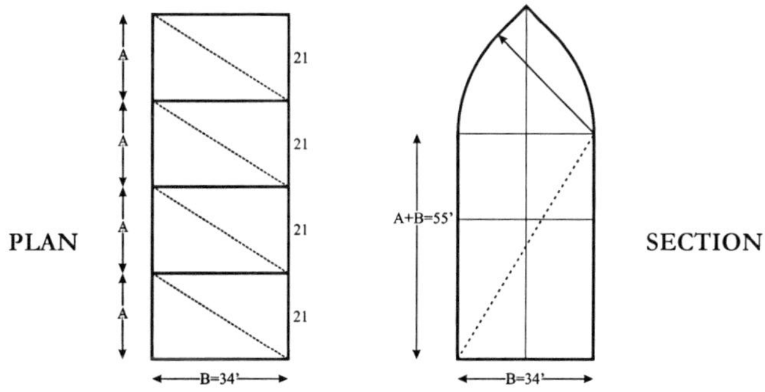

Figure 22

Analyses of this kind are two-dimensional. They are valid when applied to ground plans, formal representations extended in two dimensions only. They are also applicable to plane sections also conforming with a two-dimensional representation.

When applied to elevations or recessed sections, I find them less convincing, elevations, in particular. These are two-dimensional representations of three-dimensional entities, so an analysis involving two dimensions only cannot be expected to yield valid results. However, this being said, these analytical methods appear to have taken research a long step forward. In my mind there is no doubt whatever, that, whether the reconstructions be correct in detail or not, methods of a similar character were involved. The buildings themselves offer proof.

(3) The Master Mason's square

Dr. Morgan's research commences with a consideration of the mutual relationship of dimensions found in a number of the greater Mediaeval Churches. He then proceeds to prove that these can be derived from the properties inherent in the Master Mason's square.

So far as can be ascertained, none of these instruments survive. The example upon which he has based his investigation was taken from an incised impression, presumably full-sized, on the tomb-slab of Hugues de Libergier, the designer of the church of St. Nicaise at Rheims, long since destroyed. A likeness of Hugues himself also appeared on this monument, a very personable young man, gowned and hooded, holding a measuring rod, and a presumed model of the building.

A similar square is shown in a drawing, probably by Matthew Paris, showing Henry III, his master-mason and another functionary; one of those charming 13th century portrait groups in which each person seems to be slightly worried, and with raised forefinger admonishes everyone else.

Another example, rough but recognisable, is shown on a page of the "Pierpont Morgan" bible of about 1250, and yet another in a 16th century print showing Tycho Brahe at work.

This all goes to show that instruments of this kind, albeit comprising strange and so far unexplained characteristics, were no whims of decorative fancy, but common working tools of the times.

Hugues de Libergier's square is illustrated below:-

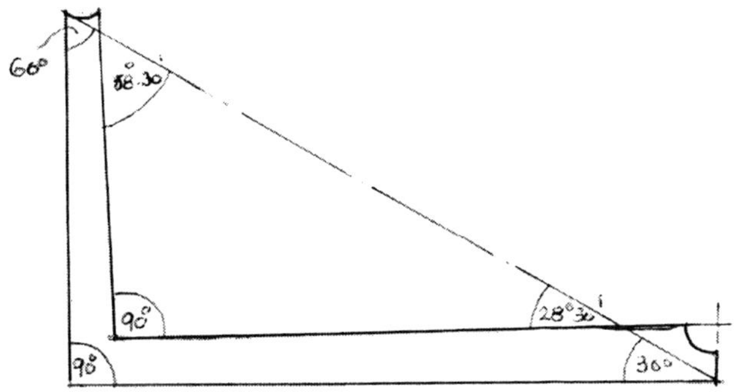

Figure 23

It will be observed that, by virtue of its tapering arms it incorporates two triangles, the 60° triangle and the Golden Section triangle previously described. No explanation can be offered for this inclusion of two triangles, so closely alike. I hold the opinion for what it is worth, that two "schools" of design existed, each adhering to a slightly different mode. The wily maker of the square increased its utility value by providing an instrument satisfying the requirements of both.

As an alternative to this, a more subtle reason can be advanced. The unaided eye plays strange tricks, so just as the Hellenic architects countered apparent sag and concavity by means of entasis, the men of the 13th century might have used an analogous process by introducing shapes generated by the triangle with the steeper sides in places where foreshortening demanded it.

Another curious feature is the scalloped ends of differing radii at the ends of the arms, Again, no rational explanation for them can be given, although it can be inferred that they are purposive. I hazard a guess that these would allow the square to be rotated round mandrels or templates with matching radii. A reason for this procedure could probably be advanced if the radii could be accurately determined, but this will have to wait on the unlikely event of the discovery of one of the original instruments in pristine condition.

Dr. Morgan deduces that in this square we have a regulating device. In the hands of the master-mason it could be used to generate and control the proportional harmonies that are so much in evidence. He also considers that its properties would render it a powerful tool in the service of structural design.

In these days we dream up shapes, and then appeal to the processes of calculation to confirm their stability. Alternatively we evolve shapes by means of calculation.

I have already indicated that I believe that the minds of the mediaeval architect-engineer moved in a totally different fashion. Experience directly acquired, and inherited knowledge including that of the nature of the materials with which they worked was allied with sensitive visual appreciation.

A shape formulated upon a deep understanding of the "rule of the square" could be guaranteed to provide a wholesome measure of stability.

I am now going to proffer some further observations in regard to the square. The long leg measures 20 $^{29}/_{64}$", which is short of the Royal cubit of Egypt by approximately 1/8". The short leg is 11 ⅓". This is about 1/2" less than the conjectural reduced foot mentioned in Chapter IV but it is almost exactly half that of a measuring unit used in ancient Jerusalem. It is also very close to one found in the Erechtheon, at Carthage and at Byblos. The two dimensions would appear to indicate a respectable antiquity for the provenance of the instrument.

Turning now to dimensions of churches culled from Dr. Morgan's work and elsewhere, my first concern is with widths. Width and height go together, but width must be determined first. It governs the nature of the crucial structural problem, the magnitude of the thrust from the high vault. Together with height it provides the shape, the frame within which the focal point, the high altar, is set. Once determined upon, it cannot be modified. A church can be lengthened, if need be, by adding more bays, but alteration to width would require that the whole structure be pulled down and rebuilt.

The width of a significant number of great churches hovers around the 34' 0" mark. Others, particularly in France and Spain, show a preferred width reasonably conforming with the 39/40 group.

The width of the presbytery of Beauvais is given as 46'0" which is very close to the addition of G.S. numbers 34 + 13 = 47. I think that considering the ambiguities inherent in our ignorance of the precise length of the foot, and the actual points from which the builders set the work out, these figures possess some significance.

When we come to lengths there are some even more remarkable coincidences:

- St. George's Chapel, Windsor, is 233'0" long;
- King's College Chapel is 289' 0" long but was *specified* as 288'0"

The length of St. George's is echoed by the Fibonacci number 233; and when we add the F. No. 55, we obtain the length of King's.

It will be seen that, once again, we are faced with a curious choice of numbers. In these instances there would have been neither practical nor particular reasons for anything but good round figures – 240 feet perhaps for Windsor and say 290 or even 300 feet for Cambridge.

But it seems that the initiating authorities thought otherwise and opted for dimensions based on the Golden Section numbers.

This ubiquitous dimension of 233' 0" or thereabouts has appeared before in Chapter V. It appears again in at least three other instances. The overall length of the Cathedral at Astorga, as 230' 0" and the width of the Cathedral of St. James of Compositely as 233' 0". The width over the crossing at Winchester is 231' 0".

The length of Salisbury Cathedral is 448' 0". The addition of the G.S. numbers 377 + 34 + 34 gives 445'0". The difference is less than 0.7%, little enough in a building of this length. Exeter provides us with 380' 0" * against 377, and York Minster 486' 0" against 377 + 55 + 55 = 487.

---

* A dimension identical with the diameters of the great ditch surrounding Stonehenge.

In two cases the dimensions are slightly shorter than that which would have been expected had the Golden Section numbers had been rigidly adhered to.

This difference could be adduced to inaccurate setting-out, or the use of the 60° triangle instead of the 58° 30' one.

These instances are but pointers, and much more detailed investigation is required. Reasonable proof of the validity of the conjectures would follow if it were found that lengths of subdivision in these structures also conformed with the numbers in the series.

In the later buildings, especially those at Windsor and Cambridge, it is possible that the dimensions were derived from Fibonacci's numbers. The earlier ones were designed before his birth, and must therefore be attributed to the direct influence of the Golden Section.

In conclusion I cite a much earlier example, built long before the commencement of the Gothic period. In Jerusalem, close by the reputed site of Gethsemane, recent excavation has revealed a large underground room. Expert opinion has pronounced in favour of this being a meeting place for the earliest Christians, probably at a time immediately following the Crucifixion. It could indeed be the earliest Christian Church. The plan of this great chamber is an almost perfect Golden Section rectangle. It measures 56'0" x 33'0".

CHAPTER XIV

THE GREAT PYRAMID

A brief assessment of the emotional and intellectual climate that prevailed in ancient Egypt must be made before discussing the Great Pyramid itself.

Professor Gordon Childe has pointed out that in earlier times there was a world of difference between the status of a tribal elder, and that of a King. The former, whatever his eminence, remained a member of the tribe, whereas a King was a being set apart and above, astride the borderline which separates humanity from divinity. In many instances, the ancient peoples believed that their gods, at times and for their own reasons, assumed physical characteristics, in human or animal form. God, invisible, is a relatively recent concept, rediscovered in comparatively recent times. If Apollo had appeared in person to a Greek peasant in pre-Socratic days, the man would have been awed and startled, but not entirely surprised, since such a visitation was accepted as a possibility within the settled order of things. We can well understand that the belief was on occasion put to nefarious use and, in parody of St. Paul, say "credulity covereth a multitude of sins".

Even the sophisticated Romans attributed something more than a hint of divinity to their Emperors, including those shabby Caesars who on occasion occupied the imperial throne.

With the spread of Christianity the cult lost its force, but still lingered in the form of a belief that Royalty was in some way representative of Divinity. As late as the 17th century, Charles I died because of an unshaken faith in the Divine Right of Kings, whereas, on the other side of the Channel, Louis XIV, he who said

"*L'etat c'est moi*" was able to maintain ascendancy (and postpone the deluge) because of it.

In our own times, we have been very close to a recurrence of primitive apotheosis. If Hitler had succeeded in his aims, I am quite sure that hordes of enthusiastic and bamboozled Teutons would have accorded him whatever 20th century equivalent there is for Godhead. It will be remembered that our own tribal leader, his purpose achieved, was rewarded by relegation to the ranks.

For the first 800 years of the ancient Egyptian united empire, it would seem that personal immortality was the prerogative of Pharaoh, and perhaps his family, alone. Those of lower station had to content themselves with the normal earthly span and it was perhaps dissatisfaction with this state of affairs which in the end caused the downfall of the Old Kingdom – the common man had come to the point of demanding his share in the good things of death. However, at the time of the building of the larger pyramids, the cult had an expectation of several hundred years, and acceptance of Pharaoh's unique privileges still held sway.

Popular imagination has adduced the construction of those great monuments to the gratification of the personal whims of megalomaniac tyrants. This is a vulgar distortion of the truth. The works were of the nature of religious offerings, and in the Egyptian mind, a counterpart of what the Gothic Cathedrals were to Mediaeval man. Whether the beliefs were true or false does not concern the issue. Pharaoh was identified with the Supreme Being and the royal pyramids combined the characteristics of a sepulchre with, when the time came, a point of departure for the God-King towards eternal habitation.

We are lucky to have a portrait of Cheops, sometimes known as Khufu, for whom the Great Pyramid was constructed. This is a tiny carving in ivory and has some affinity with the paleolithic head mentioned in Chapter V (p.27). Both have the same quality – neither are stylised – they are instinct with life. Here we have no remote hieratic image, no effigy of a half-crazed despot. Instead, we see a face that expresses wisdom and understanding. The wide distance between the eyes indicates intelligence and intellectual power, the square chin firmness of purpose, and, best of all, the set of the mouth, humour.

One can well wonder how he regarded the aura that surrounded him. He could have hardly doubted the mortality of his own flesh, and so would have concluded that his immortality was of another order altogether. Perhaps he could have said, with a great poet, who died not long ago, and whose star is presently somewhat in eclipse:

" Death opens unknown doors
It is most grand to die ".

The foregoing has been written as a brief reconstruction of the background to the task which confronted the Pyramid designers.

There is nothing of today with which it can be compared. This was no exercise in the merely mundane. It was nothing less than an affirmation of faith, in the service of an unique being, the all-powerful God-King himself, who, moreover was no passive onlooker, but according to the evidence, active in his participation in the design.

We can be certain that the circumstances demanded the most careful consideration and preparatory work, and we must pay homage to the grandeur and simplicity of the concept evolved from this.

Incidentally, the job itself combined practical ends with religious ones. During the periodic inundations, no work was possible in the fields. Construction of Pharaoh's eternal habitat provided a stop-gap alternative, whereby the unemployed could earn their subsistence allowance.

This is the moment to remark on the nature of the chosen design.

As mentioned in Chapter V (p.33) the Golden Section is implicit in the actual dimensions, and the angle of elevation.

However, this type of pyramid has other affinities. It comprises 5 planes, its base and four triangular sides, and these are separated and defined by 8 arrisses. If we postulate the Golden Section rectangular solid, in which the pyramid is, as it were, embedded, this has 5 extra sides (the sixth, the base of the pyramid, has already been counted). Again these five sides are limited by 8 arrisses. These characteristics are fundamental and inevitable, and I have little doubt they were taken into account when the choice of form was made.

For those who have not had the opportunity of seeing the monument for themselves I suppose the normal picture conjured up by the word "pyramid" is vague, comprising the structure itself, a camel or two, and some date palms, seen in silhouette against a gaudy crepuscular sky. This impression requires rectification in order to appreciate the quite staggering bulk of the work and the magnitude of the task that faced its designers and constructors.

To commence with, the base covered an area of some 14 acres, and involved setting out a perfect and level square each side of which was to be 754 feet long. This was a formidable task; it must be remembered that the available instruments and appliances were of the most rudimentary type, no steel tapes, neither tachometers nor theodolites, nothing but wooden rods, and measuring lines consisting of knotted thongs or fibrous ropes, subject to stretch and contraction.

When the builders had finished setting out, checked the opposing diagonals, and found that the square was perfect, they no doubt heaved a sigh of relief. For ultimate success depended upon absolute accuracy. Nothing less would do.

This feat, however, was but a beginning. Much greater concentration and effort was required. The problem remaining was how to set out four sloping sides to meet at an invisible point nearly 500 feet up in mid-air. The slope of each plane had to accord with the same and particular angle and to be maintained throughout construction. This angle had to be dead correct to start with, because the aiming mark, the apex, was a point in space, non-existent until it was reached.

The slightest deviation at the outset between the angle of one of the sides and its neighbours would be proportionately magnified as the work proceeded, and successively more difficult to correct, would have led in the end to an aesthetic calamity. If this had happened, it would have taken a bold man to appeal to Pharaoh's sense of humour by imploring him to look on the funny side. Pharaoh's least scathing reply could have been that, to him, the point was obscure. Nearly 5000 years later we must compliment the builders. Due to their meticulous care, there is no record of a contretemps of this sort.

XIV: The Great Pyramid

Having sketched in the design and setting-out considerations, we must now turn to a short survey of implementation.

Two different sequences were involved in the actual building operations, the construction of the main mass, the inner cores including the passageways and chambers, and the final external facing.

Limestone was employed for both purposes, and the external sheath, a little of which remains, was of a superior quality and grain, whitish in colour. Some experts hold that this was painted, others that it was unadorned. For my part, I am inclined to the latter theory, on grounds of common-sense, not only because of the enormous area involved (about 10 times the area of the above-water hull of the Queen Elizabeth II) but because of attendant application difficulties. No painter could maintain a foothold on the steep slopes, so the work would perforce have demanded cradles, arranged to slide down the sides and also move in transverse directions, when required. The rigging and manoeuvering of these would be an inordinately time-consuming job in itself, and unless the decorative paint was infinitely superior to any that we know today, the work would have been unending, repaint immediately following repaint. The dignity of the great monument would have thereby suffered the affront of this continuous assault.

There is no controversy regarding the treatment of the apex. This was either gilded or sheathed in gold. One can well imagine the awe-inspiring effect at dawn or sunset, the great monument glimmering in a violet half-light, its summit aflame.

Examination of the actual construction reveals that the main mass consists of a central pillar, surrounded by a series of concentric abutments. The diagram, Figure 24, shows this. In plan it is thus rather like a very large square onion.

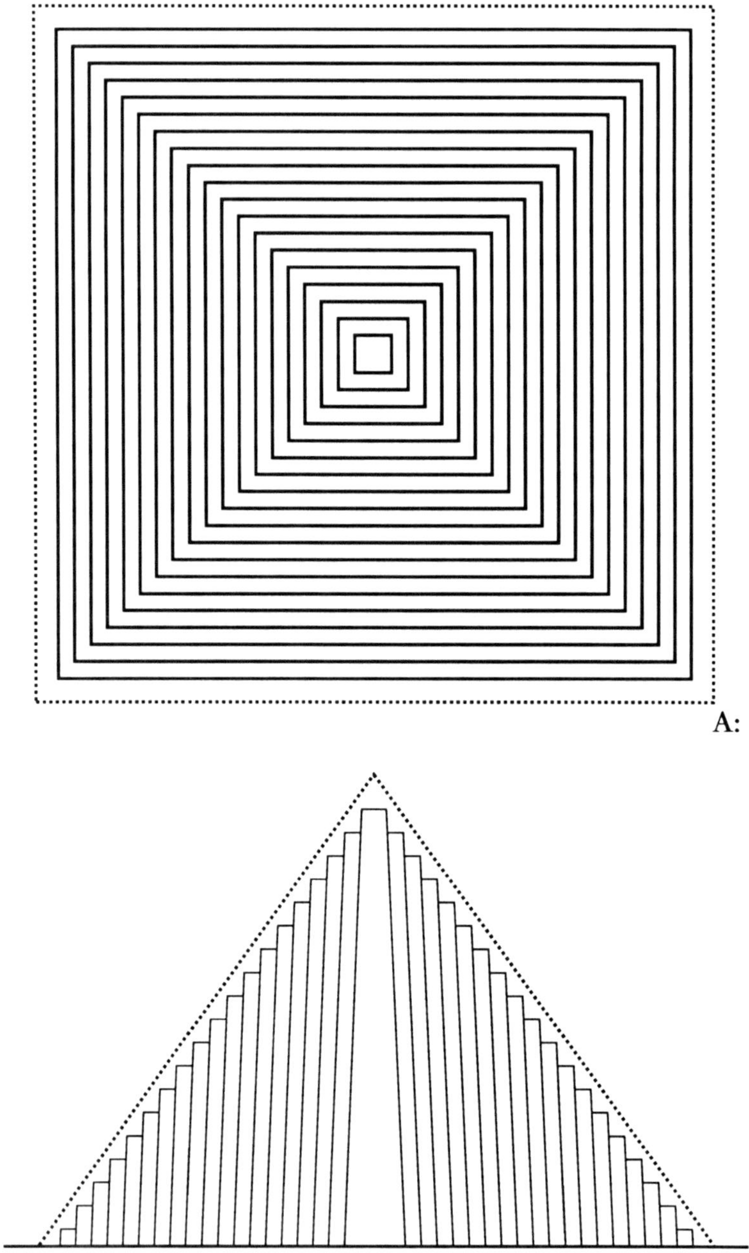

A: Plan

B: Section

Figure 24

This, at first, would appear to be a very strange technique, but a little consideration will show that it is the outcome of a very ingenious construction programme.

One might well have imagined that the easiest and simplest method would have been to build up in successive layers. This would have two major disadvantages, the first that there would be only one working platform, far too big at the commencement, diminishing to nothing at the end. The second would be of a more serious nature.

It would mean that every one of the two million massive stones would have to be cajoled up a slope of approximately 1 in 3/4, and in the last stages, the length of this haul would be a matter of some 600 feet. Now, apart from the tedious and dangerous nature of this operation, always accompanied by the possibility of disaster, it would require an enormous amount of timber for the sledges and runways, a material that was in very short supply in the Valley of the Nile.

A variation of this technique is implied in a theory that involves a temporary ramp up which the stones could be hauled. As the work progressed, this ramp would require heightening and lengthening. For its construction, it would require materials at hand, not timber, but sand, perhaps mud as a binding agent, and boulders. Assuming a gradient of 1 in 10, during the completion stages, it would have required to be about a mile long and some 500 feet high at its maximum. Allowing say a width of 20'0" for the trolleys to work and pass, and slopes of 1 in 1 ½, I calculate that about 15 million cubic yards of material when consolidated would have been tipped. This neglects the question regarding the nature of the junction between the ramp and the building. The slope of the return end of the ramp would again require to be 1 ½ to 1 and in the final stages this mass of material would have engulfed the pyramid itself. I cannot subscribe to this theory at all.

It has obviously been dreamed up by those with little practical knowledge of large-scale civil engineering works.

Herodotus visited Egypt at a time about midway between its inception and now. According to his account, he spoke with some who had access to memories of a living tradition. He says nothing about ramps, but specifically mentions cranes and sheerlegs.

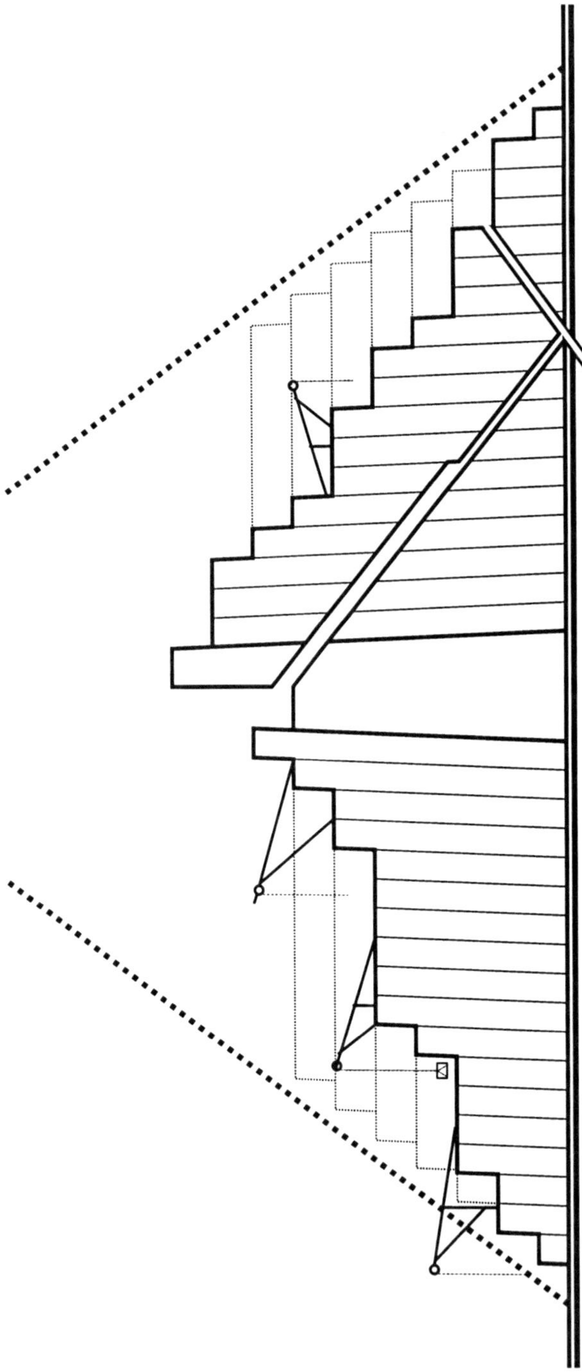

THE CONSTRUCTION OF THE GREAT PYRAMID
HYPOTHETICAL CROSS SECTION

THE ADOPTED METHOD OF BUILDING A CENTRAL CORE WITH SURROUNDING "RINGS" ALLOWS A WIDE DIVERSITY OF WORKING PLATFORMS FOR HOISTING, MANOEUVRING AND SETTING THE MILLIONS OF STONES INVOLVED.

Figure 25

I am quite unable to account for the neglect of this source of information in order to build up other and untenable theories.

If programmes of this kind had been placed before Pharaoh, I dare say that he would have commented that whereas he had all eternity before him the builders themselves hadn't.

It would seem that a much more logical scheme was evolved, It did not exclude inclined haulage altogether, but it effected a very considerable reduction and furthermore kept the lengths of haul within manageable limits.

A hypothetical stage in the construction is shown in Figure 25. It will be seen that the work can be accomplished by horizontal traversing and direct lifting. A lift of perhaps 50 feet would be well within the scope of these accomplished engineers, even if the appliances were of a primitive nature. There are a number of working platforms serving not only as stages between lifts, but allowing work to go on at various levels simultaneously. The scheme would also allow for the utmost flexibility in regard to the required level and area of any platform at any one time.

Incidentally, there were 39 steps in this colossal four-sided stairway, 19 up one side, a wide one at the top and 19 down the other (see Figure 24 B).

Whether this, or any other system was adopted, it should now be clear that the work combined the characteristics of a major engineering project with those appertaining to a severe mountaineering expedition.

To emphasise this latter aspect, it should be mentioned that a well-known climber, one whose foot had trodden the Himalayas and the Andes, lost his life on the great Pyramid. He was up there merely for the fun of it, not for the purpose of handling and setting great chunks of stone each weighing 3 tons and more.

The dangers attaching to lofty construction permeated the whole project, but in particular the work on the sheathing. The angle of the sloping sides is very great, in point of fact precipitous, and the sheathing itself smooth, offering nothing in the way of hand and foothold when setting the successive courses. Properly speaking, it was work for steeple-jacks, yet somehow or other it was undertaken and completed with the aid of rough labour, presumably that of workers from the fields and captives of war.

These men must have worked with death ever in attendance at their elbow for, one imprudent movement, one false step, and they could say, with Iras, "*The bright day is ended, and we are for the dark.*"

I must now try to deal with a paradox – the question why a great monument exhibiting a number of Golden Section characteristics should not so conform in one vital particular, namely its height.

Figure 26 shows an elevation in relation to a Golden Section rectangle:

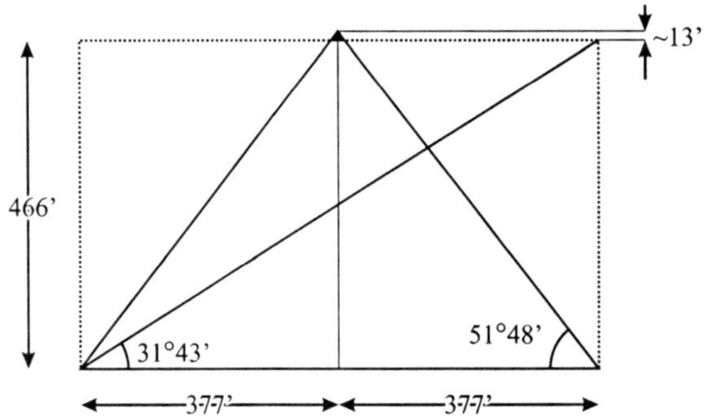

Figure 26

The apex is some 13 feet above where it ought to be, and I put forward the following reasons:

1. There was a slight mistake in setting the angle of elevation to start with, or perhaps a deviation from the correct angle during construction. Both of these alternatives are within the bounds of possibility. We have seen that the builders were meticulous in their accuracy to the point of fanaticism, but error, quite pardonable in the circumstances could have accrued from lack of sophisticated equipment.
2. In the last chapter I drew attention to the possibility that the mediaeval architects simulated Golden Section triangles by using triangles with sides of steeper inclination to compensate for parallax error. This may have well been the intention in this case, a device rendered ever more necessary by the great height concerned.

3. The third reason may be even nearer to the truth than the others. I have said that the pyramid is almost totally contained within a Golden Section "solid". Only the apex projects above. I would suggest that this had a mystical significance. The Pyramid texts make mention of "*Pharaoh rising (from his pyramid) to Heaven amidst the imperishable stars.*"

   The Golden Section parallelepiped could have symbolised the temporal and terrestrial bounds of earthly existence, and the apex the point of departure, the juncture of the visible and invisible, transcending time and space.

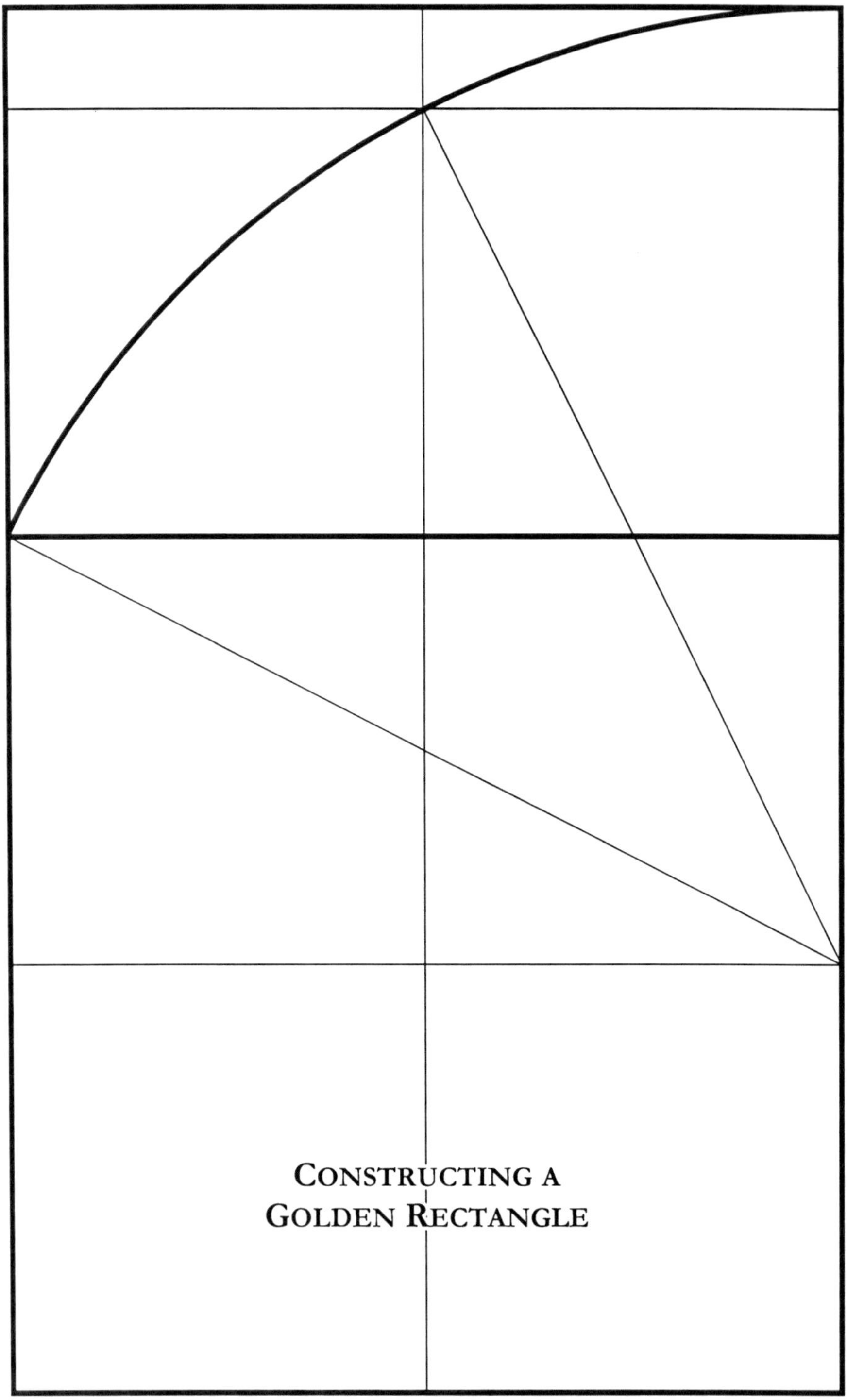

CONSTRUCTING A
GOLDEN RECTANGLE

CHAPTER XV

INSTRUMENTS

This book approaches conclusion with a study of the various instruments by means of which most of the objects made by man have been hitherto designed, set up, or set out. The three fundamentals are the line, later incorporating the measure, the trammel or compass, and the square.

The first can serve two purposes; initially as a method of establishing a straight base, in the form of an actual line, or in the form of a straight edge. In later days, its length subdivided into units of measure, such as in a measuring tape or a rule, it became the means of establishing exact dimensions. Additionally, the line with a bob or weight attached to one end can perform the duty of indicating the true terrestrial vertical.

The principal function of the second type is that of describing circles or arcs, and the last to indicate and set up the various rectangular and other angular relationships that we require to regulate our affairs.

Modern instruments for fieldwork are more sophisticated versions of the basic instruments. The theodolite measures and defines both horizontal and vertical angles to a large scale, and the tacheometer provides an optical means of measuring distances to a reasonably high degree of accuracy.

As I shall presently show, the square took over one job from the compass, that of indicating a right angle. It did not supersede the instrument, but its invention allowed this important job to be performed in a much simpler fashion.

Before we start to consider the instruments in detail another point must be made. The three types each comprise two categories, the one concerned with delineation in miniature and the other with actual implementation. That which is first indicated on paper by the use of the draughtsman's scale and set-square receives its practical enlargement to full size by means of the tape, the tachometer and the theodolite, and there is a profound difference between the usage of the set square and the working square.

It can be accepted that the invention of the square was long preceded by those of the line, the measuring rod and the compass; but it is impossible to determine which of these should be accorded priority.

The difficulty arises because of the ambiguous nature of the usages, the line as an element of radius, or a line as a line or measuring device. Let us deal with this aspect first.

(1) The Line and the Rod

Early man must have woken up to the fact that a straight line is the shortest distance between two points a long time before Euclid so defined it. I neglect great circles and the curvature of the space-time continuum. These concepts had no part in the mental processes of homo when he first acquired sapiens. It would not have been long before he realized that certain physical dimensions were applicable to all sorts of needs, and could be stabilised for use in the form of wands or laths of a set length. For work on a large scale, and to save time, these standard lengths could be incorporated in the line by means of knots or marks of some other kind. A line so marked, or a wand would be put to use according to the task to be performed, the setting-out of earthen walls to an enclosure, or the proper determination of the size of a burial place. I have already indicated the nature of the measuring devices that could have been evolved, and one can imagine for example that work on a passage grave would be preceded by the ceremonial unwrapping of the sacred wand, perhaps a canne in length, so that the width of the excavation could be prescribed.

Neither the line nor the measuring rod would be the slightest assistance when work was on a small scale and it may well be that this would be carried out without reference to measurements.

Domestic utensils, personal adornments and such can all be made by eye, yet I have a suspicion that measuring devices for helping in the design of these commodities were not entirely absent.

That "tally-stick" previously described, with its curious arrangement of 55 inscribed notches may well have served this very purpose. The notched bodkin dug out of the same hole could have combined a measure and an implement. We find the same sort of thing today. The spare edges of my slide rule are marked off in inches and centimetres, so that measurement can if required, follow calculation. I seem to remember a paper-knife, the blade of which was treated in the same way; and set-squares that could be used for measuring purposes as well.

(2) The Compass

Early compasses, or rather, trammels are believed to have consisted of a length of bone or hardwood in which two pointed flints were inserted at a fixed distance apart, one as a centre, and the other as a scribing point. This device incorporated the characteristics of both instrument and tool, for we can deduce that prehistoric man did not conform with our method of designing on paper to serve as a guide for execution. He worked direct – *premier coup* every time – and the intricate patterns on the ancient stones of Ireland bear witness to the acquisition of a high degree of proficiency.

However, an instrument of this kind had two disadvantages. The radius was immutable and it was quite unsuited for work of any kind except that on a small or moderate scale.

So I surmise that the first compasses were of a different nature and comprised a line with a peg at one end and an animal at the other. Natural circular forms are very rare. The ripples from a pebble dropped in a pool, the disc of the Sun and that of the Moon at the full are the only examples that come to mind. We know that both of the latter were subjects of veneration by early man, and indeed continued to be so well into historic times.

Animals, the hooves of which left crescent-shaped imprints similar to the shape displayed by the young Moon, the Maiden, were held to be associated with that deity so it can be taken that a discovery that a tethered beast could crop out a simulacrum of both

Sun and Moon must have made a potent impression, a supposition which tempts me to offer a new and totally spurious derivation for the origin of the inn sign "The Goat and Compasses". The device by which this effect was obtained would thus be held in high esteem, greatly magnified when it was found that a human being could himself produce the same, merely by substituting himself for the animal, thus inaugurating the first and noblest of geometric figures.

This primitive apparatus, centre-peg, line and marking peg initiating so many early tectonic shapes has accompanied us out of the abyss of time, and still retains its useful functions. It has the advantage of providing an infinite variety of radii and we still put it to use today; especially those of us who are gardeners.

It may have been employed once on a very grand scale indeed. A magnificent curved arched brick viaduct carries the railway line for some distance between Surbiton and Hampton Court Stations.

The legend persists that setting out on the ground presented such difficulties that the engineers were flummoxed, until the problem was finally resolved by an intelligent ganger fixing a centre, and pegging out the radius at the end of a long line attached thereto.

(3) Squares

It is perhaps impossible to isolate the origin of the psychological need to produce rectangles. These rarely occur naturally and when they do, it is seemingly by accident. The reasons, whatever they are, may have something to do with those that impelled evolving hominids to assume an erect posture, in which case there is a kinship between the rectangle and the plumb line.

The origin of the first square must forever remain obscure, but we can be certain that it involved the compass, unless someone arrived at the concept by the unlikely process of suspending a plumb line over that progenitor of the spirit level, an unruffled pool. Or perhaps some prehistoric forerunner of Pythagoras conducted an experiment that generated the first right-angle.

We can imagine him running into the family shelter shouting "Mum, Mum, come and see what I've drawed." And the long-suffering Paleolithic mother would follow him to where a strange diagram had been inscribed in the sand. Lacking the slightest glimmer of comprehension she would yet be lost in fond admiration and moreover somewhat relieved in that she could forego the usual objugation hitherto: "You nasty little boy – scrape it out at once"

We can reproduce this experiment for ourselves in accordance with the diagram and procedure shown below.

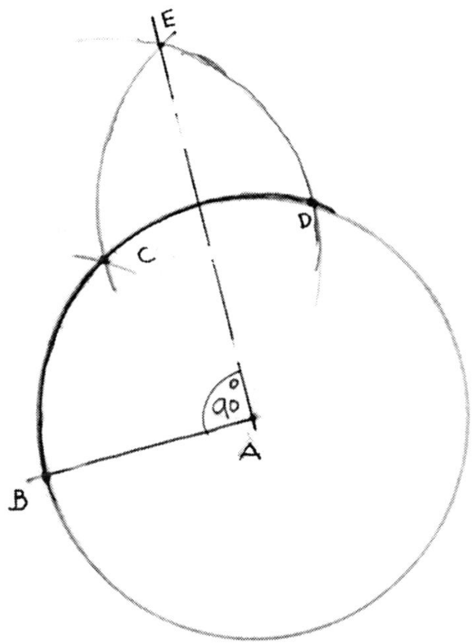

Figure 27

1. Describe circle with radius AB
2. Inscribe Line AB cutting circumference at B
3. Using the same radius cut circumference successively at C and D.
4. Still using the same or any greater radius describe arcs CE and DE intersecting at E
5. Join E and A. The line EA is at right angles to AB

N.B. There is a simpler method still, but this would involve an extension of A-B, and the use of a different radius to arrive at point E

Whether this historical reconstruction be correct or not the square was somehow evolved, and the instrument came into being, adopting various forms.

I must emphasise once more, that there are two general types.

The older is the working square, its very much younger brother being the set-square, the instrument that simulates the stipulated tasks, usually to a scale less than full size. These were made in wood or metal, until celluloid and other plastic materials took over. They take the form of triangles, the most common being that containing the right-angle and two of 45° in which base and perpendicular are of equal length, and that containing 90°, 60°, and 30° where for obvious reasons, the length of base and perpendicular differ.

The working square, used *inter alia* by masons and carpenters, is concerned with the actual implementation of the job. It takes a quite different form. It comprises two arms or blades joined to meet at an angle of 90°. These members differ in length. The mediaeval mason's square, copiously described elsewhere in this book is of this type. Had the intended use of this instrument been confined solely to that of setting out or checking right angles, the arms could have been of equal length, but the fact that it does not so conform implies that other usages were involved.

Mediaeval pictures show the instrument in use for the mundane purpose of squaring stones. One, however, slowing an operative thus employed also depicts the approach of the master mason clad in hood and gown. The facial expressions displayed by master and man seems to indicate that trouble is brewing: "How many more times have I to tell you not to use my square for – etc."

Two less venerable examples merit consideration. Some time ago at Beauvais, a gang of masons were lapidating the church of St. Etienne. Like many another French church, this is a mixture of styles ranging from Romanesque through late Gothic to a neo-classical "West End" and Tower. Our Gallic friends had no inhibitions about this sort of thing, and the results that they sometimes achieved prompt me to wish that Hawksmoor had followed suit when he completed the twin towers of Westminster Abbey, instead of opting for a somewhat heavy and insensitive 18th Century attempt at Gothic.

At Beauvais, the work concerned the insertion of new stones into the fabric of the walls to replace those that had decayed.

A truck would deliver a load of lovely honey-coloured material, and the masons then marked the blocks to the appropriate size. Then the stones were re-loaded for sawing at a nearby yard. My wife told me that the foreman was brandishing a fine steel square, so I set out in the hope that this might possibly conform to the dimensions of that belonging to Hugues de Libergier. In the event, it did not, but in some ways it was just as interesting as if it had.

At first, what with my indifferent French and the foreman's thick accent, the entente was anything but cordiale, but by dint of gesticulation and phrases such as "architect Anglais" and "étudiant des bâtiments Gothiques", communication was at length established. Even then there was some difficulty in explaining that it was his square that I wished to measure, not the stones. Scratched, notched and dented as I found it, it was a noble instrument, still showing the gleam of bright steel and I shall later give my reasons for believing it to have been of some considerable age.

After I had taken the measurements, the foreman and I parted with mutual expressions of good will. As I looked back, he was engaged in a whispered conversation with one of his assistants, at the same time tapping his forehead in a significant manner. Intercepting my glance, his native Gallic courtesy prevailed and he desisted.

The length of the arms measured on the outside proved to be 604mm 429mm respectively:

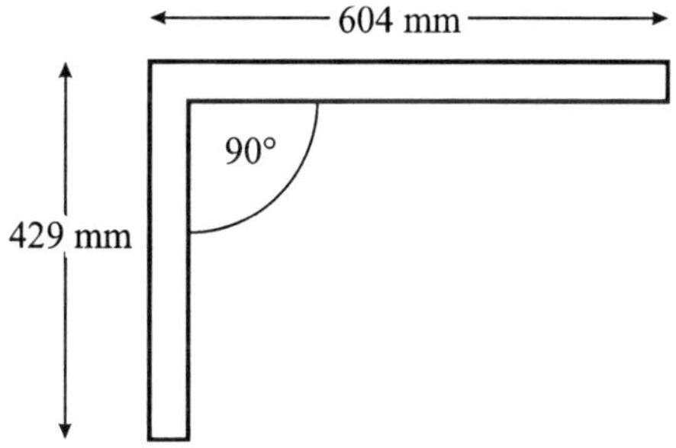

Figure 28

and dividing the greater dimension by the smaller we arrive at a proportion of 1:1.408. The relationship of the diagonal of a square to its sides is ~1:1.414, so it is evident that the length of the two arms were intended to express this function (not the ratio between tritone and tonic, even if the relationship is even closer).

The second example, a wooden carpenters square came into my possession some years ago. A very long-established family concern occupied a shop and workshop in a back street in the neighbourhood where I then lived. The business consisted of joinery, repairs to furniture and upholstery, but failing to keep pace with the onrush of the 20th Century, it had fallen into decay. I was not surprised to see that it was closing down, and that the poor remnants of stock and tools were for sale. I bought a bundle of well-worn tools including a wooden square.

The elderly woman who sold them to me told me that they had belonged to her deceased husband and before him to her father who had inherited them from his father. The square is an ordinary carpenters square; the arms measure 22 ½" and 14" on the outside. Dividing one length by the other, we arrive at 1:1.618 which is so near to the Golden Section proportion that the difference is visually indistinguishable.

In addition to this, the inside measurements of the arms, within 1/8 of an inch, conform with those measured on the outside of the master masons square in use at Rheims more than 700 years ago.

These two examples are sufficient to indicate that a whole new field of study remains to be explored. Both squares exhibit functional characteristics in addition to that of indicating a rectitude of 90°. The present owner of the first example would be unlikely to make use of the additional property. As anyone who has tried knows, there are practical difficulties in establishing true rectangularity to a small scale by use of the diagonal. In this case there would be no need. The instrument itself was there for the particular purpose.

In regard to the second example, I think it unlikely that a humble mid-19th century craftsman would be familiar with the properties of the Golden Section as such, although in his daily work he might, unwittingly perhaps, make use of them.

To my mind, both examples point to the thread of continuity upon which this book lays so much emphasis.

Let us consider the dimensional aspect. If you or I were to set about making a square we would tend to commence by marking out one side using familiar dimensional units, in these days calling upon Imperial measure, and in the future the metric. In other words, so many round inches, or aggregations of millimetres. In the case of the French instrument, it will be noted that the two dimensions conform with no known measurement unit, neither metric, nor even that within the system that preceded it. This leads me to the supposition that this square, perhaps of no very great antiquity itself was based on an earlier prototype.

The length of the long arm of Hugues de Libergier's instrument approximates closely to the Royal cubit, and the short one to the postulated reduced foot. The length of the perpendicular in the example we are considering appears to comprise two of these shorter units, or something very close.

The longer blade of the carpenters square also seems to indicate that two units were involved, each of 11.25".

This might have emanated from a different source, such as were once in use in Phoenicia, but I think it more likely to have stemmed from an essay in inaccurate vernacular of the kind mentioned in Chapter VI.

Do not misinterpret my meaning. I am not implying that the makers, in this case (or the preceding one) consciously adopted two antique units, even though they might have been aware that they were perpetuating on the one hand the Golden Section ratio, and on the other the relationship of the diagonal to side.

By rights we should have found the long arm of the carpenter's square to be two feet in length. This could have been done with a minimal increase in the amount of material used in the two members.

We do not find this, but are confronted instead with non-conformity in respect of the measuring units current at the time of manufacture.

This on the face of it, is very strange, and one can only surmise that the constructed shape of these instruments followed a tradition reaching very far back into the past.

Even the modern commercially produced set-squares seem to exemplify this lingering tradition. All the so-called 12"/45° set squares that I have measured give a dimension something short of the current foot. The equivalent seems to lie between 11.80" and 11.825", that is within a figurative hairs-breadth of the "short" foot.

With the advent of the draughting machine, these instruments are now becoming obsolete. The draughting machine itself has no great life-expectation and will no doubt soon lie in the next stratum of the rubbish dump; for, I envisage the draughtsman of the near future, wreathed in cigar smoke, supplies of nutritive and sedative tablets ready to hand, reclining in his swivel chair, the while he manipulates an array of press-button controls.

CHAPTER XVI

LARES ET PENATES

This chapter could have well been entitled "Fun with a Foot-rule" for it was this unassuming instrument that aided a selective eye in all the preliminary work. Various personal possessions were subjected to measurement, and a consistency of Proportion revealed itself.

Had the analysis of the results been regarded as an end in itself, rather than a step in an enquiry, it could have been set down in tabular form, perhaps interesting to some extent, but demonstrating that on occasion brevity can constitute the very essence of the vacuous. This sort of thing would have been very close to a process of measuring, weighing and tabulating Turner's watercolours and assessing their merit therefrom; an exercise patently absurd, but, I suggest, not unrepresentative of much that goes on nowadays under the title of "research".

The point that I wish to make is that the discovery of an identity of proportion between objects and articles otherwise apparently unrelated prompted further study and consideration of the things themselves.

The chapter is therefore a kind of epitome of the whole book, a series of short essays that studies not only proportional relationships, but something else, other and quite different, the characteristics and history of the diverse objects that demonstrate this affinity of proportion.

To begin with, it may be as well to consider the relative strength of the factors that control our acquisitive desires, for the purpose neglecting those of usage – value and cost.

These apart, I believe the most potent to be shape and colour and of the two, colour exercises by far the greater influence. Even when it comes to the choice of food, or the packages within which much of it is now contained, it is the bright and often discordant hues that attract the customer's eye and fills the vendor's pocket.

To point to an example, I am quite sure that the sales of confectionery, no matter how delicious would drop sharply if the items were just grey or black. The reaction is of course instinctive. For reasons not yet understood the eye and mind perceive colour and the contrasts between hues much more readily than those between one shape and another. Road signs, especially those of a mandatory character are enclosed within the confines of standardised shapes, but nevertheless depend for their effectiveness upon brilliant colour. Military uniforms once made use of bright hues to aid instant identification until it became evident that there were certain disadvantages in providing the enemy gratuitously with the same information.

And, it may be asserted, in the arena, it is the colour of the matador's muleta rather than its cut that invites the bull to death.*

I suppose the basic reason is that essentially, we and all other sentient creatures are 'children of light'. Without light there can be no such thing as colour.

This being so, the attention paid to shape by our forebears both recent and remote is all the more remarkable, and what is stranger still, a predilection amounting almost to an obsession for a particular rectangular shape, or relationship of parts, generated by the Golden Section.

As I have already emphasised in Chapter III, it would seem that this choice of a controlling shape was at some times contrived but at others instinctive. Both conscious and subconscious selection played

---

* This may be incorrect as it is now believed that these animals possess limited appreciation of colour values.

their part, and still do so. Many of us see a current example in the first category most days of the week when we look at the double arrow, the colophon adopted by British Rail, contained within a Golden Section rectangle.

To come now to the various articles examined, I have a little to say about the 4th century wine jar mentioned previously. It measures 10" × 6 ¼". A long time ago it would have offered solace to its possessor, but, alas, now no longer, because the excavator's pick punched a large hole in its base.

The carpenter's square has already been described and is only mentioned again because a square of this kind could have been used to determine the proportions of another article made about the same time. This is a mirror set in a "Tunbridgeware" frame. Tunbridgeware is a term applied to the decoration of surfaces by applying sections of veneer of contrasting, grain and colour alongside one another to form patterned margins and the like. This example is not amongst the finest. The maker was not only hampered by a limited selection of material, but was evidently no great adept in fitting the pieces together. This mirror measures 17½" x 28½", or a proportion of 1:1.63.

More satisfying in every way is a little clock made by Marc of Paris. This has been demurely nibbling away at eternity for about 150 years and will no doubt continue to do so for as long again.

The inlaid wooden case is 5¼" wide × 8⅜" high and the centre of the dial lies on the intersection of the diagonals of a square of unitary dimension 1 in relation to the total height of 1.62.

Descriptions of the remainder of the articles so far identified now follow, the oldest being accorded precedence –

1. Demeter

Some 2300 years ago, this little baked clay head was in all likelihood an object of veneration, hanging upon some humble wall beside the Aegean. It now hangs on an equally humble wall in the South of England, and if not actually venerated, is nevertheless regarded with affection.

The goddess's features have suffered the assault of the centuries, worn smooth by the touch of countless fingers since she first emerged from the kiln. In fact, her mouth has practically disappeared altogether, and at the risk of committing two-fold *lèse majesté*, I have to confess that now her profile bears a distinct resemblance to that of Queen Victoria expressing extremes of non-amusement.

The little thing measures 3⅛" x 1⅞" – 1:1.62.

2. A Portrait of Charles II

Almost every age has produced a number of artists of the highest calibre together with a rout of those much less gifted, and in Western Europe, there have been periods during which one or more of the Muses have exulted, while the others nodded or slept.

This was the case in the early 13th century, for whereas architecture and sculpture often attained to near perfection, graphic and pictorial art rested. at a recognisably lower level.

I refer in particular to the sphere of character delineation. At Rheims, Chartres and elsewhere, the superb sculpture is instinct with life. Work on vellum during the same period, despite vigorous design and delicacy of execution is generally lacking in this respect. We observe the groups of little figures stereotyped in attitude and gesture, the faces all alike, their features expressing little joy or sorrow, neither benevolence nor anger. It all appears to lie within a mannerist context, a submission to the rule of an accepted formula.

The reason for this defect, if defect it is, must be assigned to the fact that the work was executed by men occupying a world quite different from those who manipulated mallet, punch and chisel. The implements involved were the quill and the brush, the material not harsh and unyielding stone, but smooth surfaces prepared to receive fine-ground pigments.

The prime purpose of the executants was the inscription of the word together with embellishment of initial letter or complete page.

Such an occupation within the peaceful ambience of cell or cloister would inculcate painstaking and meticulous workmanship, but hardly a passionate involvement in the tensions and stresses of human life.

XVI: Lares et Penates

The illustrations, of which I speak are, as a rule, found on the pages of manuscripts prepared by craftsmen, themselves learned, for the use and edification of the still more learned. These books were not intended for, and indeed, would have been unintelligible to the greater part of the community, intelligent, no doubt, but illiterate.

For those, the promulgation of a religious or historic lesson (usually one and the same) had to be carried by means of stained glass. Here we are confronted with a curious paradox, for Mediaeval stained glass combines unsurpassed colour harmonies, superb compositional design and at the same time the naive attitude towards delineation of animal and human form that we find in the manuscripts.

It is perhaps by conscious intent, symbolic rather than naturalistic. For had the cartoons been prepared by a Velasquez or a Frans Hals, somehow called back from later times, displaying these artists' subtle physiological insight, it is highly probable that Mediaeval man would have made neither head nor tail of the finished products.

There was one difficulty inherent in this method of presentation. If all the faces were alike, all imprints from one subjective stencil, how was the most important person in a group to be differentiated from the remainder?

The solution was simple and direct. It was the portrayal of him or her (generally him) as being much the largest figure in the composition.

This device was inherited from remote antiquity. I am sure that it was never intended to be taken literally, but it may be assumed that the intent, by dint of long usage, misfired in the minds of unsophisticated onlookers, where oft times no comparison between portrayal and actual subject was possible.

In ancient Egypt, Pharaoh was usually depicted as being of gigantic stature; yet when a current Pharaoh appeared in public, the ordinary man in the street would not be misled, despite the double crown, into a belief that he was more than say 5' 4" in height. The true physical height would be known and accepted and on a par with the knowledge that given the temerity and a good hard pull, Queen Hatsheput's formal beard would part company with her chin.

This pictorial association in which physical size stands in for spiritual intellectual or social dominance lingered for a very long time in a different fashion.

I have a contemporary engraving of Charles II executed by William Sherwin, a friend of Prince Rupert, to whom he dedicated a series of prints, showing the Monarch and his consort.

Charles was a very tall man, even by today's standards. In this engraving his head, feet and hands are shown as being disproportionally small. By means of this device, a very tall man is given super-human stature.

The attributes of pomp are lacking, but not those of Royalty, for despite the plain, even sober attire he stands (taking size of head, feet and hands as a guide) a veritable King, every 108 inches of him.

The effect is enhanced not only by the quality of the three-hundred year old glass through which the print is seen, but even more so by the shape of the panel within which the portrait is set 7.7" x 12.4" = 1:1.61.

3. The Tapestry Screen

We possess a large folding screen composed of tapestry on a wooden frame. The tapestry depicts a hunting scene, or rather a pause for refreshment before resuming the rigours of the chase.

The elegantly garbed participants, careless of their finery recline in a leafy dell, busily tucking in, and in the middle distance, late-comers to the feast are seen to be urging their steeds in order to arrive before the last delicacies disappear. The scene is idyllic, although the artist, possibly a pupil of Lancret has not failed to indicate tensions within the group. A young man is paying much attention to his attractive companion. Another lady, wife or discarded mistress perhaps, watches the couple with disfavour, even to the neglect of the tasty morsel on her two-pronged fork. The line of her gaze, striking across the picture, seems to be the axis about which the whole composition has been arranged. Another matter of interest manifests itself, for there is a curious affinity between this tableau and one that appears on the so-called "Standard of Ur" which also depicts a convivial scene, a drinking party.

I do not refer to the fact that in both, flounced skirts are worn, in the first, of silk by the ladies and in the other, cut from felt, by an all-male assembly of the bald-headed. The bond of similarity that links both is this: In each case, on the extreme left and in the same position relative to the main group, a servitor, seen back view, kneels to replenish supplies. In the 18th century group he wears breeches, gaiters and a tricorn hat. In that from the 3rd millennium B.C. he is attired in the same fashion as his superiors but there is little to be seen beyond a plump flounced posterior managing to express extreme concentration on the job in hand. The Standard of Ur, discovered and painstakingly restored under the direction of Sir Leonard Woolley is an historical record of the first importance. It was made with solemn intent, and the tableaux bring before our eyes, in peace and in conflict a way of life forgotten these 5000 years.

Yet, amidst the pomp and ceremony, this touch of undoubtedly intentional humour, speaks to us across the ages.

To return to the screen. I am of the opinion that the tapestry was woven in the early 19th century, possibly at Beauvais where, upon the restoration of the monarchy, the royal looms resumed work and produced reproductions from the original 18th century cartoons.

The screen is in four sections and measures overall 5' 3¾" high by 8' 6" long – that is 5 × 8 "*pieds du roi*", a perfect Golden Section rectangle. Apart from this characteristic, it is of interest that the use of an archaic obsolete measuring unit persisted into relatively modern times, even if the reason therefor was the obvious one of working to a size and pattern established some seventy years earlier.

However, it instances before our very eyes the last link in a chain of tradition, forged moreover within the confines of a geographical area where the roots of tradition strike very deep.

4. The Lutar

From the early days of the 16th century until the latter part of the 17th, Western Europe, and England in particular appear to have gone music-mad.

We have ample evidence that monarchs and nobles alike, the lower orders and even architects succumbed to the prevailing fashion.

It is on record that Inigo Jones himself, bearing the first arch-lute to be brought into the country was for a time incarcerated in Dover Castle; for King James' port security officials suspected that this strange looking affair was some kind of infernal machine, a threat to the monarch's safety or even his life.

We are told that barbers' establishments provided instruments for the entertainment of those awaiting shaving or cupping.

Not here, the penetrating squeal of rebec or treble viol nor the rude, not to say downright disgusting noises from the blown tube. The softer voices of cittern or bandurria were deemed more appropriate, and less likely to put the practitioner off his stroke at a critical moment.

During this period, the queen amongst plucked string instruments was undoubtedly the lute, an instrument of oriental origin, believed to have arrived in the West in company with returning crusaders, and if this be so, to have taken some 300 years to acquire perfection.

In its time, it was esteemed in much the same way as the grand piano in the 19th and early 20th century, and maintained its pre-eminence in this respect for some 200 years or more.

However, it is extremely difficult to play. Even a moderate level of accomplishment demands a deal of ability and submission to an arduous discipline over a period of years.

To this factor that led to its eventual decline, another must be added. In its heyday it was equipped with 12 or 14 strings, termed courses, some of which were doubled, so the left hand had at most 7 finger positions to deal with. Over the years, in an effort to achieve greater range and sonority, the number of strings was greatly increased. This made the instrument still harder to play and much more difficult to tune and to keep in tune.

Towards the end, the current gibe was that ten minutes playing was preceded by at least a two hours tuning. The problem was exacerbated by the nature of the strings themselves. Modern strings, machine-made from synthetic materials, each conforming to a particular pitch still include occasional "rogue" strings. In the 16th and 17th centuries, the strings were made from natural gut. They were supplied in bundles containing strings of various thicknesses and the instrumentalist had to select by eye not only for correct pitch, but for quality as well.

There must have been a deal of exasperated comment when strings broke before attaining the required tension, or buzzed or clattered when they did.

Bearing these troubles in mind, it is not surprising to find that the instrument, facing a vast improvement in keyboard instruments, such as harpsichord and spinet, strung more manageably with wire, and easier to play, should decline into a phase of total obsolescence, to last for nearly two hundred years.

There were two exceptions. In the first place, the Neapolitan mandolin, a somewhat insignificant member of the lute family, strung with wire and played with a plectrum. This, save for similarity of shape has nothing in common with the parent instrument, neither range, expressiveness, nor nobility of tone. The other is a simplified lute, known as the lutar.

I have been so far unable to trace its origins and whether in fact, it is a direct descendant of the classical lute, or a revival.

Certain characteristics point to the former hypothesis, and I should mention that I found a near sister to the instrument that I possess, hanging on the wall of a mountain-hut in the Tyrol. It was in shocking condition, several segments of the belly and two strings being missing, yet it was still playable. All traces of varnish had long since disappeared and it was black with age and the traces of the countless grimy fingers that had handled it.

As its name implies, the lutar is a marriage between lute and guitar. It has six strings and a relatively narrow fingerboard, but here the resemblance ends. The body is the same shape as that of the lute, rather like a pear cut in half and it has the same elaborately carved decorated rose, the sound hole. The belly comprises the same number of segments as the lute, that is, eleven, and the peg-box terminates in a square-topped "Milanese" scroll much resembling those of late 17th century lutes.

The tone quality also differs from that of the guitar. The voice of the latter is rich and sombre, with a hint of depths wherein it is not always wise to stray. That of the lute or lutar may be better described as golden, providing an acoustical counterpart of the effect of morning sunlight upon water.

It is as if the instruments each in their fashion represent the respective dominions of Dionysus and Apollo.

As one well-known guitarist said, after playing on mine "It's such a civilised instrument!"

It also conforms with its ancestor in another way. The table of fine straight grained pine is but $1/16^{th}$ of an inch thick, and the belly, the joint between each segment reinforced with parchment, little more than 1/32". The instrument is quite sizeable, in fact, over three feet long yet it weighs but a shade over 31oz.

My lutar is fairly new, probably made in some small central European factory. I first saw it hanging in a shop window. It was the beginning of a love-affair that has lasted for more than 20 years.

We come at last to the crown of the matter, the proportional characteristics. The length of the body is 21" and the overall width 13". The discovery of this conformity with Golden Section dimensions was satisfying enough but when I came to measure the overall length I was even more delighted. From the top of the scroll to the lower end of the body it is a little more than 39" long.

CHAPTER XVII

EPILOGUE

Our mode of thinking is determined largely by the numerical system that we use and have grown up with.

For a very long time past, we have thought and lived within the decimal ambit; indeed, in its earliest form this could have been the first basis of numerical application. Man had ten fingers and ten toes before he possessed much else. Long after, we still find it natural to think in tens, and multiples of ten.

Another system is that based on twelve instead of ten. This has no natural origin, no physical prototype but, as mentioned previously, was held in high regard some 4,000 years and more ago. We still make some use of it dividing (if for but a little while longer) a foot into twelve inches and the day into 24 hours. We multiply five basic units by twelve and assemble these components to form respectively the minute and the hour. The minds of those who once used this system exclusively must have worked in a fashion differing from our own. If they had been presented with the current aspects of decimal application it is unlikely that they would have been able to come to terms with it.

The sparse traces of a third method exist, that based on the use of 2 as a multiplier – 2, 4, 8, 16 etc. This has a noble simplicity, but judging from what remains, it could never have been prevalent. It appears to have been confined to capacity measure, and in an isolated instance, to the measure of weight, the 16 ounces that comprise the pound.

From the evidence that has led me to write this book, I postulate a fourth system, or rather, a mode comprising the numbers generated by the Golden Section, a likelihood strongly reinforced by the diversity of context within which these numbers have been found.

This mode would differ from the others inasmuch as the usage seems to have been in large measure esoteric. When it is mentioned at all, it appears as a subject for academic speculation, or within the confines of aesthetic theory. Without currency in the service of the multitude, it has nevertheless exercised the most profound of influences. We can well conjecture that those whose hands it guided, even if not bound to maintain and guard a "mystery", would have little reason to comment upon it. As something ingrained in the mind it would have been, in a sense, a commonplace, a familiar discipline, calling for no special mention within or without the ambit of their daily preoccupation.

This is the probable reason why something so powerful and so far-reaching in its implications has been largely forgotten save for sporadic use. Those who were wholly involved are long since dead, leaving indeed their works and some of their tools and instruments to contemplate, but nothing of their methods save frail traces, some of which have been gathered together in these pages.

It remains to consider how we, in this present century, can repair the discontinuity and build anew upon such ill-defined foundations.

That this is possible, I now propose to show.

It has truly been said that knowledge of the past begets wisdom. In other words, the more we know about former times, the better equipped we should be to regulate our current affairs. This truism has been recognised more or less clearly throughout the ages, but at present it takes on an aspect of special urgency.

Archaeological investigation in one form or another is our surest method of recovering the past. It is not only a quest for factual evidence but also a means of providing a measure of contact between ourselves and our forebears. It marches with strictly scientific procedures, and joins hands with art in all its manifestations, with history, and with history as-yet unwritten. This is as it should be, for who amongst us would not wish to know more of their own ancestors, a hundred or a thousand generations back?

XVII: Epilogue

Of late, the new and powerful tools provided by applied science are allowing us to probe ever more deeply, and with increasing certitude. Yet another could be provided by an extension of the enquiry that has been outlined in this book, moreover one calling for no expensive equipment. An observant eye, a few simple devices for measurement, and the time to exercise them would constitute the sum of the requirements.

As I have mentioned, the usage of Golden Section characteristics has been by no means universal, and herein lies advantage; for this very nonconformity can enable us to identify the thread leading back into the labyrinth.

Field investigation would be followed by the sifting of the gathered evidence from prehistoric times onwards, layer by layer. The identification of proportional, numerical or mensural conformities, or one with another, between megaliths on the Atlantic seaboard, stone circles in Eire and earthworks in Wiltshire could reveal a pattern of distribution well worth further investigation. The enquiry would not be confined to the permanent and the indigenous. It could also be applied to artefacts, both in regard to their own nature and also the locations in which they were discovered.

Things from more recent times could come under review, culminating in a study of the most copious source of all, the Mediaeval period, somewhat after the fashion indicated in Chapter XIII.

Research on these lines would be fascinating in itself and, in addition, a means of forging an historical chain, link interlocking with link.

In another sphere altogether, I suggest that those who design structures could find that economic and aesthetic advantages could accrue from the exploitation of Golden Section characteristics, just as it did in the Middle Ages.

The results could point to methods in themselves possessing the additional merit of simplified design procedure, the computer meanwhile assisting.

The adoption of such a discipline could even result in a feedback modifying the outlook of the practitioners in such a fashion that we might yet see a F.I.C.E. on a site, emulating Master Simon:

> "In magistral manner here and there setting out the work already conceived in his mind by the yard-stick of his eyes."

In a third realm, astronomy, the application of the principles is likely to be far less rewarding. I have already remarked that there appears to be no close correlation between Golden Section characteristics and certain similarities found in Nature. I do not say that they are entirely unconnected, because both may stem from the same Origin, but it would seem unavailing if not downright foolish to try and force an inbuilt human slant upon nature in general.

I have mentioned an attempt to equate the proportional distances of the planets from the Sun with the Fibonacci series. This is but little more convincing than Bode's "Law". My own contribution to this dubious concept is to establish a reason why Pluto, otherwise an outsider, can be brought into the picture, if such a picture can be said to exist. There is some reason to postulate the existence of another planet more distant than Pluto, and much more massive, so notwithstanding doubts and reservations I plunge even more deeply and suggest that attention might be concentrated for awhile on a region defined by the surface area of a sphere with a radius of 54-56 AU struck from the centre of the Sun.

In conclusion, I return to the underlying theme, to which this book owes its title; the long years since the emergence of humanity, and the unimaginable span that preceded this.

I believe that the steps that led to the appearance of Man as a sentient self-conscious creature took place on this Earth. There has been time enough, and there is no need to postulate incursions from other worlds. Yet in fairness, and taking into account the immensity that lies behind us, we should not entirely reject this possibility.

If the prime basis of scientific exploration is not erroneous, light from sources in outer space has been speeding towards us at 186,000 miles per second for some 3,000 million years. This elapse is minimal, first because it represents only the time for the signal to travel, not the duration of the source, and secondly, this inordinate stretch is believed to represent, but the third of a total, the limits of which are set by physical impediments to cognition.

Notwithstanding this, advanced thinkers consider that the time-distance antagonism, at present inhibiting journeys beyond the solar system, will be eventually resolved.

If there are grounds for this belief, it follows that the same problem could have received successful attention elsewhere, and at other times.

The concept involves abolition of what we choose to call time-flow. However, as each light particle, in a sense, carries its own quantum of experience, the flux of time would seem to require not abolition, but reversal.

Thoughts of this nature prompt some appreciation of the magnitude of the past. Much could have happened and been forgotten within such a period, by comparison with which the 10,000 years or so over which semi-certitude regarding our own history extends pales to insignificance.

Perhaps the last word on the subject, the Alpha and the Omega of it, is contained in the Christian credo which, laying stress on an event of supreme import, states that this took place "before time began".

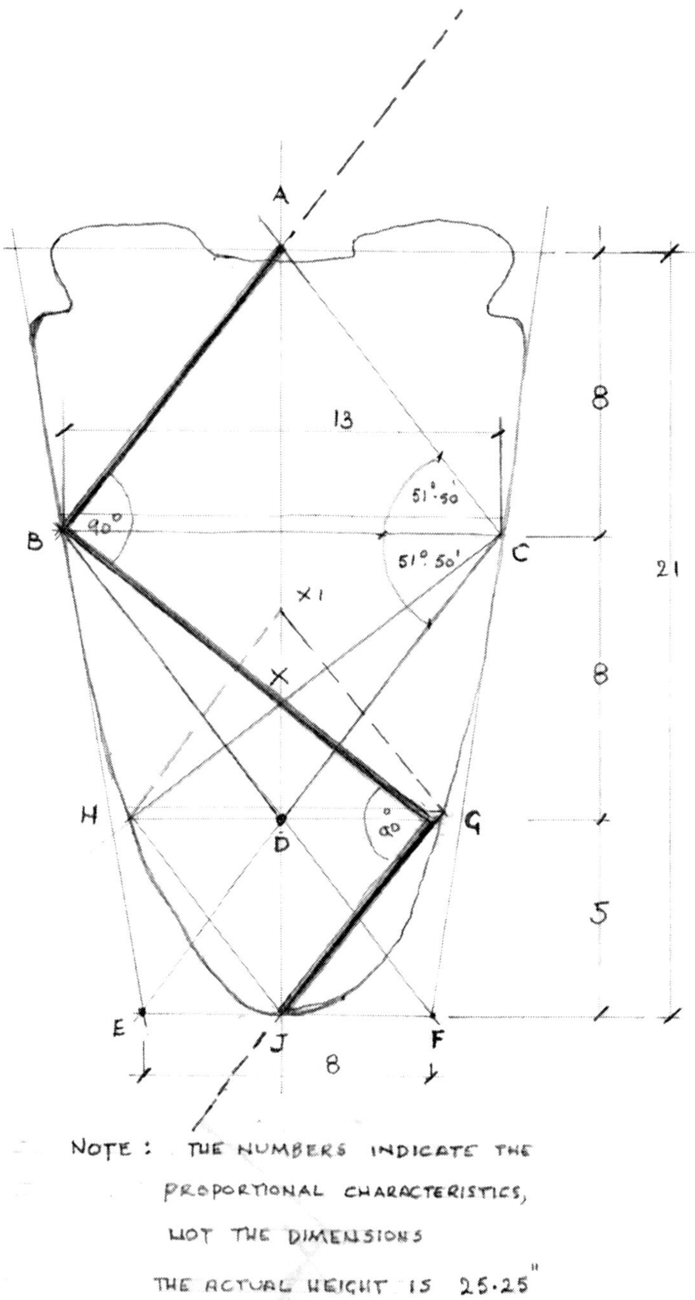
A
B
C
D
E
F
G
H
J
13
8
8
5
21
8
90°
90°
51°50'
51°50'
NOTE: THE NUMBERS INDICATE THE
PROPORTIONAL CHARACTERISTICS,
NOT THE DIMENSIONS
THE ACTUAL HEIGHT IS 25·25"

Figure 29

APPENDIX A

THE PALETTE OF NARMER

The unification of the Kingdoms of Upper and Lower Egypt to form one realm was achieved about 3200 B.C. The first Pharoah was either King Narmer or King Menes, but it is uncertain whether these were separate individuals or whether the two names denoted the same person.

Figure 29 (opposite) shows a commemorative shield-shaped "Palette" of hard slate, carved in low relief some 5000 years ago. The superb artistry and craftmanship displayed in this work is evidence of a tradition long preceding its manufacture. In this instance, although the work is to a small scale, it could have been of great importance, being commissioned by, or on behalf of the supreme authority, and when the artist received his instructions he was no doubt briefed in regard to the subject matter and the overall dimensions; and not only these, but the context of proportion within which he was to work.

There is a certain relationship between this and the design of King's College Chapel, Cambridge. In both cases it can be inferred that the properties of an accepted proportional canon were known to the commissioning authority as well as those whose job it was to execute the work. In other words, there was tacit agreement in advance. But in each case it would appear that within certain limits the artist was free to depart from a rigid geometrical context, *if this led to an enhancement of the aesthetic effect.*

The diagram shows my conjectural reconstruction and analysis of the methods adopted in the setting out.

Do not credit me with the subtlety and simplicity which marks the conception. These must be attributed to a great artist whose name we shall never know.

The design shown is that of the obverse which is divided horizontally into three sections, the upper two being identical in height, and the bottom one somewhat less. The heights are in the proportions 8:8:5. The common base B-C of the two upper portions is 13 in the same proportion.

Triangle ABC is a triangle contained within a Golden Section rectangle having sides in the proportion of 8:13. Triangle BCD is the same, but reversed. The lowermost triangle DEF is similar to triangle ABC, characteristically the same but within a rectangle with the proportions 5:8. The sum of the perpendiculars of these three triangles is 21.

Each triangle is similar to the projected elevation of the Great Pyramid, constructed about 800 years later.

It will be noted that the angle of inclination from point J determines the length of the edge of the lower major division H-G.

A-B and G-J are of course parallel. B-G is at right angles to both. Although the primary rhythm is provided by two large triangles forming a lozenge and a smaller pendant triangle, these in their turn generate two more pleasing shapes ABXC and XHJG and a further rhythmic emphasis. This last is indicated by the heavier delineation of A-B, B-G and G-J, and it will be observed that the configuration is analogous to the trace of the Fibonacci numbers described in Chapter VIII and shown in Figure 13 which, it will be recalled, is not unlike the behaviour of the bob of a free pendulum as it loses momentum. It can be extrapolated in either direction, but in a downward one it crosses and recrosses the vertical axis at rapidly diminishing intervals.

In Egyptian art, and long after, in Ancient Greece, Greece, the crossing of two members, generally legs or arms, is a salient characteristic. In this example, the lower intermediate panel is occupied by two leonine chimera, which taken together comprise the major element in the pictorial composition. The long necks of these creatures cross twice, once at point X, where B-G and its complementary C-H intersect, and again at the apex of triangle HGX1, an exact repetition of triangle DEF below and a mirror image of the identical reversed triangle HJG.

The total height of the plaque is 25¼". This conforms with no unit of measurement contemporary with the work.

A measure of about 25" has been inferred from evidence found in Persia, Greece and Palestine, but this is all much later in date, as is the 12th Dynasty cubit of 26" - 27".

The "Great U" of 40" is far too large and the 21" cubit too small.

However, dividing the height by two we get 12.625" and if we again divide this into the postulated reduced inches (digits) (20.62 / 21) we get a figure of 12.88, very close to the Fibonacci number 13. The total height of the plaque, using 13 reduced inches, would work out at 25.48" an increase of less than 0.25%. This is well within the possible margin of error that the material and the technology of the time could impose.

We can therefore conjecture that if the plaque was to be of a certain non-conforming size, the preliminary setting out could have been done to half-scale, using a recognised unit, and then doubled in scale for implementation.

I offer this explanation as a not unreasonable conjecture.

Enough has been said to show that the Golden Section forms the basis. It is patently manifest that this design is neither the result of chance, nor intuitive approach, but of conscious intent.

It is of no great importance whether the unit of measurement mentioned above, or any other unit, was adopted.

What is of note is the fact that the concept of number had to be invoked, either to set up triangles, or to define the rectangles that enclose them, and that the numbers involved are once again found to conform to the Golden Section series.

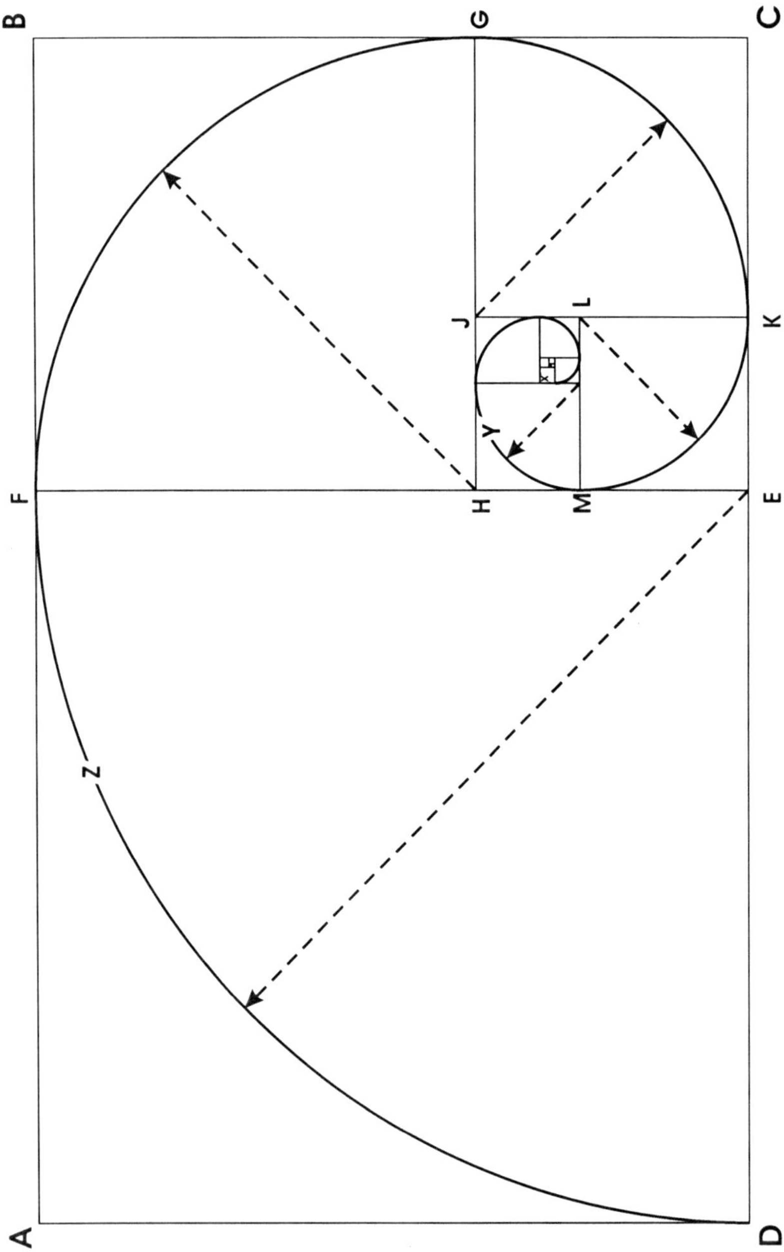

Figure 30: Approximation of a Golden Spiral, using arcs

APPENDIX B

1. The accompanying diagram and description shows how the Golden Section can be used to generate a spiral form, the volute. This affinity between the characteristics of the two has often been remarked upon. It occurs in nature, particularly in shell-forms, those of the nautilus, and the fossilised remains of its predecessors known under the name of ammonites.

I can see little connection between the shape of the nautilus and that of the man-made volute, beyond the fact that they both possess spiral characteristics. The former, beautiful indeed, are close-knit, the spiral expanding but slowly. The volute founded on the Golden Section proportion expands rapidly. This must be expected. At the lower levels, the difference between the numerical steps is undetectable in visual form even if, as we know, they are successive incremental steps proceeding from the infinitesimal.

Once they arrive within the sphere of graphical cognition, they increase in an orderly fashion by leaps and bounds, but at a rate which soon takes them beyond the range of visual comprehension. They, like ourselves, come out of the dark. For a short while they manifest their growing stature and then vanish in the direction of their ultimate goal.

This process of continuous acceleration (the term used in its strictly technical sense) makes it almost impossible to display adequately in graphic form the illustration of the idea that was promised in Chap. IX.

A demonstration to correct scale would require a sheet of paper of such a size as to place the publication of this book beyond the range of possibility.

However, I shall do the best I can within the allowable limits.

2. I said that the repetition of the sequences at increasing numerical levels appear to come full circle. That the sequential items assuming various values return to their origin, but at an ever greater magnitude, thereupon resuming a set numerical course.

I also said that the term "circle" was in the nature of a figure of speech, and that the phenomenon could better receive graphic delineation by assuming that these chosen initiatory points lay on the trace of a spiral curve.

For reasons just now given, distance between these points cannot be readily plotted to scale, but the idea can be grasped by inspection of diagram Figure 30 where readings Nos. 1, 20 and 44 or, for that matter, any others such as 45, 64 and 88 are indicated at points marked X, Y and Z.

APPENDIX B

3. The Construction of the Volute

1. Set up G.S. rectangle ABCD, and cut this by line EF, at a point where AE is to AB as 1:1.62. The original rectangle is now divided into a square AEFD, and a further G.S. rectangle EBCF.
2. Treat this rectangle in the same fashion to provide square EBGH, and another consequent G.S. rectangle HGCF. This in turn generates square JGCK and G.S. rectangle JKFH. The process is repeated until the resulting squares and rectangles become too tiny to draw.
3. a) With point F as centre, strike arc DE with radius FE.

   b) From point H, strike arc EG with radius HE.

   c) From point J, strike GK, with radius JG and so on, to continue the conformation as far as may be possible.
4. The construction brings out another fundamental difference between the shape of the man-made volute and those evinced by the natural forms referred to at the beginning.

   The volute as drawn does not evolve continuously, insofar as each successive portion of the curve is drawn to a different radius.

   Arc DE is the same radius throughout its length, EG, also circular, depends upon a smaller one. There is therefore a change of characteristic at the junction between the two at E.

So far as I can ascertain, natural forms are of variable curvature throughout. Nature's geometry is evidently of a different order depending neither upon man's units of measure, nor his use of the compass.

Pattern based on the Fibonacci Series
(*see also* p.101)

Dramatis Personae

Absolon, Dr Karl
Achilles
Agamemnon
Apollo
Arthur
Aristarchus of Samos
Avignon, the Master of

Bartók
Beethoven
Bode, Johann
Bonington, Richard Parkes
Brahe, Tycho
Brown, G. Spencer

Caligula
Cantor, Geog Charles
Cherubini
Childe, Prof. Gordon
Cheops, see Khufu
Churchill
Claudius, Imperator
Copernicus

Debussy, Claude A.
Des Pres, Josquin
de Manchicourt, Pierre
Devil, the
Dunstaple, John
Dürer, Albrecht

Edward I
Eudoxus
Euclid

Fibonacci (Leonard of Pisa)

Gibbs
Gildas
Gracie, M.S.

Harvey, J.D.
Helmholtz, Hermann von
Henry III
Henry VI
Heracles
Herodotus
Hoyle, Prof.
Hugues de Libergier, Master

Incitatus (a horse)

Khufu

Leonardo da Vinci
Louis XIV
Lund, Frederick Macody

Mantegna, Andrea
Marcus Aurelius
Margyas
Menelaus
Merlin
Michael, the Archangel
Morgan, Dr. Bernard George
Mozart, W.

Narmer, or Menes
Newman, Barnet
Nono, Luigi

Index

A

B

Index

H

I

J

K

L

Index

O

P

Q

R

Index

S

T

U

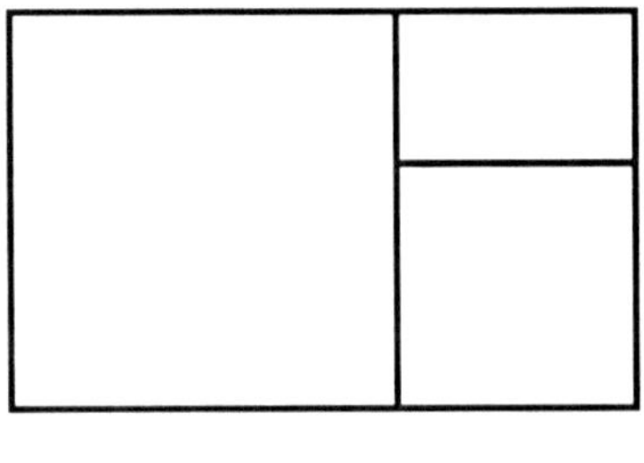

Printed in the United States
116583LV00001B/58/P

9 780955 706813